In the Middle of the Fight

An Assessment of Medium-Armored Forces in Past Military Operations

David E. Johnson • Adam Grissom • Olga Oliker

Prepared for the United States Army

Approved for public release; distribution unlimited

The research described in this report was sponsored by the United States Army under Contract No. W74V8H-06-C-0001.

Library of Congress Cataloging-in-Publication Data

Johnson, David E., 1950 Oct. 16–
In the middle of the fight : an assessment of medium-armored forces in past military operations / David E. Johnson, Adam Grissom, Olga Oliker.
p. cm.
Includes bibliographical references.
ISBN 978-0-8330-4413-6 (pbk. : alk. paper)
1. United States. Army—Armored troops. 2. United States. Army—Armored troops—History. I. Grissom, Adam. II. Oliker, Olga. III. Title.

UA30.J62 2008
358'.18—dc22

2008029010

Cover photos courtesy of the U.S. Army Center of Military History

Published 2008 by the RAND Corporation
1776 Main Street, P.O. Box 2138, Santa Monica, CA 90407-2138
1200 South Hayes Street, Arlington, VA 22202-5050
4570 Fifth Avenue, Suite 600, Pittsburgh, PA 15213-2665
RAND URL: http://www.rand.org/
To order RAND documents or to obtain additional information, contact
Distribution Services: Telephone: (310) 451-7002;
Fax: (310) 451-6915; Email: order@rand.org

Preface

The U.S. Army is in the midst of a major restructuring and transformation effort to prepare itself for the challenges of the 21st century. Its ultimate objective is to create a campaign-quality army with joint and expeditionary capabilities. As part of its transformation, the U.S. Army is fielding medium-armored forces, the Stryker brigade combat teams (SBCTs), to give the current force increased capability. Medium-armored forces are also central to the U.S. Army's vision of the Future Force, with the Future Combat Systems–equipped brigade combat teams considered an important component of that force.

This report presents a historical analysis of how medium-armored forces have performed across the range of military operations since World War I. Its purpose is to help inform U.S. Army decisions about the Future Force.

This research was sponsored by the U.S. Army and conducted within the RAND Arroyo Center's Strategy, Doctrine, and Resources Program. RAND Arroyo Center, part of the RAND Corporation, is a federally funded research and development center sponsored by the U.S. Army.

Because this study was initiated prior to Fiscal Year 2002, there is no Project Unique Identification Code (PUIC) associated with this project.

For more information on RAND Arroyo Center, contact the Director of Operations (telephone 310-393-0411, extension 6419; fax 310-451-6952; email Marcy_Agmon@rand.org), or visit Arroyo's Web site at http://www.rand.org/ard/.

Contents

Figure and Tables

Figure

Tables

Summary

The purpose of this study is to draw insights about medium-armored forces from past operations to help inform decisions about U.S. Army transformation and the design of the Future Force. The study is a qualitative assessment of the employment of medium-armored forces in the 20th century and it relies on a multicase, comparative historical approach. We assess U.S. and foreign experiences to analyze how medium forces performed across the range of military operations in complex terrain and against different types of opponents, as shown in Table S.1.

Tasks

The project's sponsor specified three central questions for the study:

- What unique capabilities have medium-armored forces brought to past conflicts, and where along the spectrum of operations have they been most valuable?
- How have medium-armored forces performed in complex terrain in the past?
- What advantages has the rapid deployment capability of medium-armored forces provided to operational commanders in the past?

Methodology

We used historical research, mainly as supplied in secondary sources, to select and develop the cases studied in this report. At the sponsor's request, we assessed each case from several perspectives:

Table S.1
Case Studies

Case	Complex Terrain	Point in the Range of Military Operations	Type of Operation	Types of Armored Vehicles and Other Forces
Armored warfare during the Spanish Civil War (1936–1939)	Broken and mountainous; urban	High	Major operations (civil war with external support to both sides)	German and Italian medium-armored vs. Soviet Union heavy
U.S. armored divisions in France and Germany during World War II (1944–1945)	Urban; hedgerows; forests	High	Major operations	U.S. medium-armored vs. German heavy
Armored cavalry and mechanized infantry in Vietnam (1965–1972)	Jungle	High	Major operations; counterinsurgency operations	U.S. medium-armored and heavy vs. Viet Cong and North Vietnamese light
Soviet airborne operations in Prague, Czechoslovakia (1968)	Urban	Middle	Strike (regime change)	Soviet Union medium-armored and heavy vs. Czechoslovakian light (mainly civilian forces)
South Africa in Angola (1975–1988)	Close; undeveloped infrastructure	Middle	Major operations; raids	South African medium-armored vs. Angolan heavy
Soviet Union in Afghanistan (1979–1989)	Urban; mountains; undeveloped infrastructure	Middle	Strike (regime change); counterinsurgency operations	Soviet Union medium-armored and heavy vs. Afghan light
Operation Just Cause, Panama (1989)	Urban	Middle	Strike (regime change)	U.S. medium-armored vs. Panamanian medium and light

Table S.1—Continued

Case	Complex Terrain	Point in the Range of Military Operations	Type of Operation	Types of Armored Vehicles and Other Forces
1st Marine Division light armored vehicles (LAVs) in Operations Desert Shield and Desert Storm, Southwest Asia (1990–1991)	Desert; limited visibility	High	Major operations	U.S. (Marine Corps) medium-armored vs. Iraqi heavy and medium
Task Force Ranger in Mogadishu, Somalia (1993)	Urban	Low	Raid	U.S. light and coalition (Malaysian and Pakistani) medium-armored vs. Somali light
Russia in Chechnya I (1994–1996) and II (1999–2001)	Urban; mountains	Middle	Counterinsurgency operations; combating terrorism	Russian medium-armored and heavy vs. Chechen light
Australia and New Zealand in East Timor (1999–2000)	Urban; jungle; undeveloped infrastructure	Low	Peace operations	Australian and New Zealand medium-armored vs. rebel light
SBCTs in Operation Iraqi Freedom (OIF) (2003–2005)	Urban	Middle	Counterinsurgency operations; combating terrorism	U.S. medium-armored vs. indigenous Iraqi and foreign fighter light

- How were medium-armored forces employed, and why does this monograph define them as "medium-armored"?
- What doctrine, organization, training, materiel, leadership and education, personnel, and facilities (DOTMLPF) insights emerge?[1]
- What are the battlefield operating system (maneuver, fire support, air defense, command and control, intelligence, mobility/countermobility/survivability, and combat service support) implications?
- Which operational characteristics that the U.S. Army expects of a transformed force (i.e., responsiveness, deployability, agility, versatility, lethality, survivability, and sustainability) surfaced, or did not?
- How did the medium-armored force under examination perform in the case environments (i.e., complex terrain)?
- What key insights emerge?

Finally, the sponsor asked us to describe any overarching insights that are common among cases.

Key Findings

Several cases examined in this study show the critical difference that even small numbers of medium-armored forces can make, particularly in augmenting light forces or when operating independently in raids or strikes. In Somalia, Malaysian and Pakistani armor provided the protected mobility and firepower required to extricate cutoff elements of Task Force Ranger. Similarly, U.S. medium-armored forces in Panama during Operation Just Cause provided a needed edge to light forces, and even the modest number of deployed M551 Sheridans provided an important capability at crucial moments in the early stages of the campaign. U.S. Marine Corps LAV units were an important economy-of-force and reconnaissance element during Operation Desert Storm. Medium-armored forces gave Australia and New Zealand the capability

[1] The facilities aspects of DOTMLPF are not addressed in this monograph, however.

to range widely and rapidly across East Timor with protected mobility sufficient to meet the threat. Furthermore, SBCTs were able to provide rapid response across a large operational area in Iraq, with greater survivability than light forces. Finally, medium-armored forces are more able to operate in areas with less-developed infrastructure. This was the case in Panama, where M551 Sheridans could cross bridges that could not support U.S. main battle tanks.

Having the capacity to rapidly deploy medium-armored forces (by air or sea) may be an important national capability. This was apparent in operations by the South African Army (SAA) in Angola and in the Australian response to East Timor. Rapidly deployable medium-armored forces were also an important capability in the coup de main operations conducted by the Soviet Union in Czechoslovakia and Afghanistan. Currently, the U.S. Army does not have a forced-entry, medium-armor capability. Although the air-droppable M551 Sheridan armored reconnaissance vehicle provided this capability in Panama, the vehicle has since been retired from the inventory. Stryker medium-armored vehicles are not air-droppable and, with their add-on armor, can only be deployed by C-17 or C-5 transport aircraft. This likely limits their movement by air to any but secure locations.

Medium-armored forces highlight the fundamental defense-planning challenge of balancing predictability and adaptability. Peacetime choices about future capabilities, rooted in judgments about likely adversaries and environments, matter greatly because most wars are "come as you are" in many respects. Medium-armored forces have experienced the majority of their difficulties when conditions on the ground differed significantly from the predictions used to prepare those forces. This phenomenon is most apparent in the case of the U.S. Army in World War II, when U.S. medium-armored forces were obliged by strategic and operational circumstances to directly engage German heavy-armored forces that possessed significant survivability and lethality advantages. U.S. Army doctrine had explicitly rejected this contingency, and this conceptual error resulted in unnecessary losses for many U.S. armor units. Similarly, while SAA medium-armored forces enjoyed great success against Angola's Soviet-supplied heavy forces in the late 1980s, the unexpected arrival of heavy armor on South Africa's doorstep led Pretoria to hedge against a future recurrence by fielding its own heavy forces.

The primary implication of this study is that the development of the U.S. Army's Future Force should be framed by a broad conceptual paradigm that embraces the complexity and diversity of the types of military operations that the nation may call upon that force to execute. In future conflict environments, the U.S. Army may face—as it has in the past—adversaries who operate in complex terrain and are equipped with heavy armor and highly lethal weaponry. In some circumstances, therefore, the materiel employed by U.S. medium-armored forces will be inherently less survivable and less lethal than the materiel fielded by their adversaries. Even if digitally enhanced situational awareness lives up to expectations, such circumstances will be very challenging, and medium-armored forces will need to compensate with sophisticated combined-arms tactics that exploit enduring U.S. advantages in artillery and air support (as did U.S. Marine Corps LAV units during Desert Storm and U.S. Army forces during World War II).

Given the breadth of cases examined in this study, we can draw an even more pointed conclusion: Medium-weight forces are useful only when deployed under one or more of the following conditions:

- by air in a way that preempts an effective enemy response (as in Czechoslovakia and Afghanistan)
- against an enemy who lacks the capability to deal with any mobile armor (as in Panama, Somalia, and East Timor)
- in circumstances where other friendly assets—e.g., close air support, artillery, a significant training differential—offset enemy capabilities (as in Desert Shield and Desert Storm, Angola, and OIF).

In short, this monograph suggests that medium-weight armor enjoys only four clear advantages over heavy armor: rapid deployability (particularly with air-droppable vehicles), speed over roads, trafficability in infrastructure not suited to heavy armor, and lower logistical demands. It furthermore suggests that these advantages are exploitable only in conditions where the resulting diminution of combat power can be accepted or compensated for by other means. Because the U.S. Army cannot expect all future operations to occur in such circumstances, it would be prudent to maintain a mix of heavy, medium-armored, and light forces that can be task-organized and employed in conditions that best match their attributes. Medium-armored forces have much to offer in such a mix.

Acknowledgments

The authors thank the many individuals who contributed their time and intellectual energy to the evolution of this monograph in its present form.

The initial support for this project, and the formative questions that guided the research, were provided by Vernon M. Bettencourt, Jr.

We want to thank our RAND colleagues, John Gordon IV, Jerry M. Sollinger, and Peter Wilson, who offered valuable advice about the ideas examined herein. William H. Taft V made invaluable contributions to the early framing of the monograph when he was a research assistant at RAND. Thomas L. McNaugher and Laurinda L. Zeman provided the resources and encouragement necessary to finish this project. Finally, Jefferson Marquis and Richard Sinnreich provided very thoughtful and useful reviews of draft versions.

This monograph is better for all of their efforts.

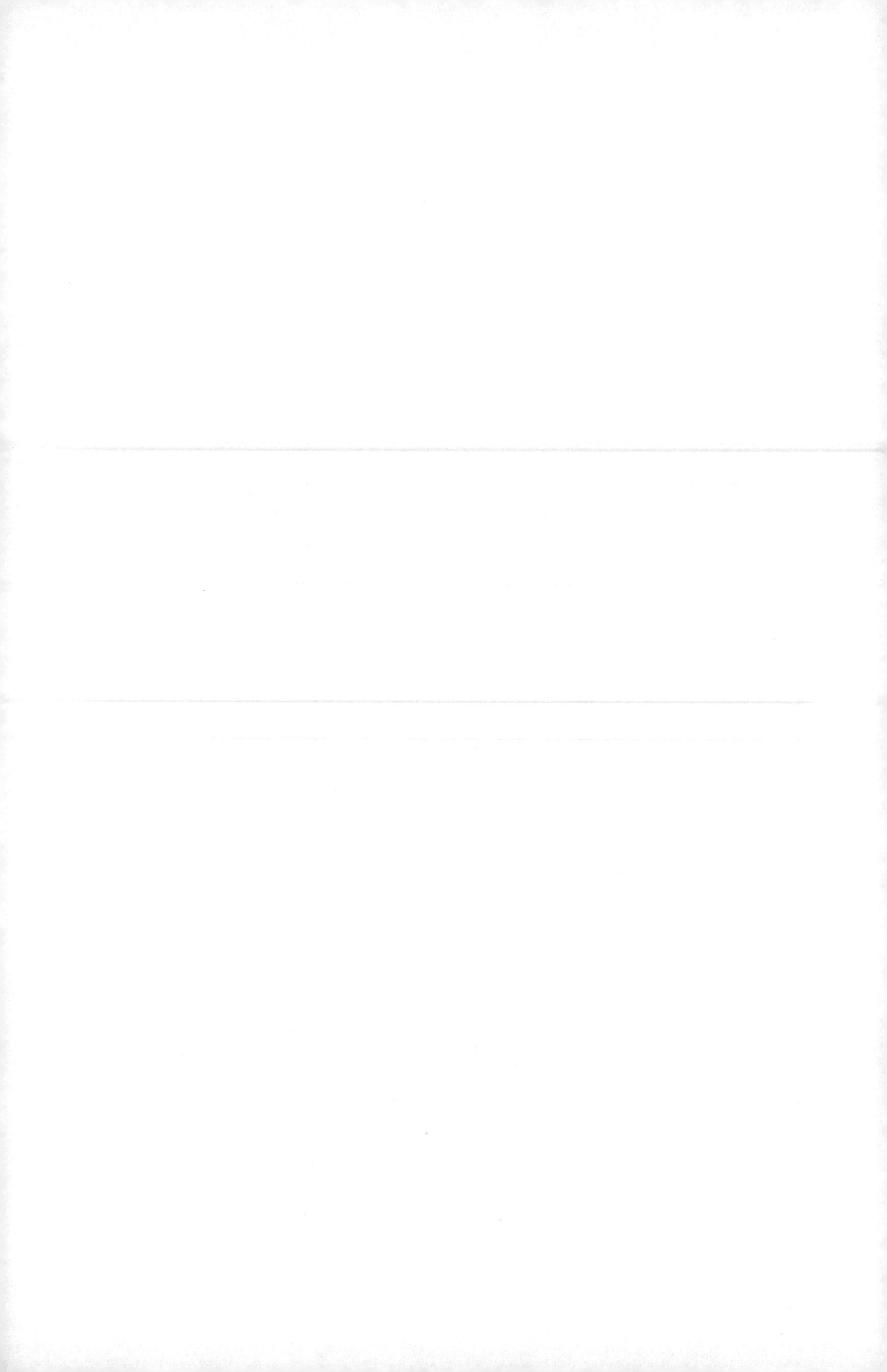

Abbreviations

4ID	4th Infantry Division
AAV	amphibious assault vehicle
ABCS	Army Battlefield Command System
ACAV	armored cavalry assault vehicle
ADF	Australian Defence Force
AFV	armored fighting vehicle
AGL	automatic grenade launcher
AH	attack helicopter
ANC	African National Congress
AO	area of operations
APC	armored personnel carrier
ARFORGEN	Army Forces Generation
ASLAV	Australian light-armoured vehicle
ASLAV-A	Australian light-armoured vehicle–ambulance
ASLAV-C	Australian light-armoured vehicle–command
ASLAV-F	Australian light-armoured vehicle–fitters

ASLAV-P	Australian light-armoured vehicle–personnel carrier
ASLAV-R	Australian light-armoured vehicle–recovery
ASLAV-S	Australian light-armoured vehicle–surveillance
AT	antitank
ATGM	antitank guided missile
ATGW	antitank guided weapon
BCSS	Battlefield Command Support System
BCT	brigade combat team
BOS	battlefield operating system
BRDM	*Boyevaya Razvedyuatel'naya Dozornaya Meshina*
C2	command and control
C3	command, control, and communications
C4ISR	command, control, communications, computers, intelligence, surveillance, and reconnaissance
CALL	Center for Army Lessons Learned
CCA	Combat Command A
CCB	Combat Command B
CCR	Combat Command Reserve
CENTCOM	U.S. Central Command
CONARC	Continental Army Command
CONUS	continental United States
CSLA	Czechoslovak Peoples Army
CSS	combat service support

DoD	Department of Defense
DOTMLPF	doctrine, organization, training, materiel, leadership and education, personnel, and facilities
DRA	Democratic Republic of Afghanistan
EW	electronic warfare
FBCB2	Force XXI Battle Command Brigade and Below
FCS	Future Combat Systems
FM	field manual
FSB	*Federalnaya Sluzhba Bezopasnosti*
GPS	global positioning system
HMMWV	high-mobility multipurpose wheeled vehicle
HLVTOL	heavy-lift vertical-takeoff and -landing
IDP	internally displaced person
IED	improvised explosive device
IFV	infantry fighting vehicle
INTERFET	International Force East Timor
IPB	intelligence preparation of the battlespace
JFC	joint force commander
JSOTF	joint special operations task force
JTFSO	Joint Task Force South
KGB	*Komityet Gosudarstvennoy Bezopasnosti*
LAV	light-armored vehicle
LAV-AT	light-armored vehicle–antitank

LAV-C2	light-armored vehicle–command and control
LAV-L	light-armored vehicle–logistics
LAV-M	light-armored vehicle–mortar
LAV-R	light-armored vehicle–recovery
LOC	line of communications
MACOV	Mechanized and Armored Combat Operations in Vietnam
MACV	Military Assistance Command, Vietnam
MANPADS	man-portable air defense system
MBT	main battle tank
MEB	Marine expeditionary brigade
MoD	Ministry of Defense
MOUT	military operations in urban terrain/military operations on urbanized terrain
MPLA	Movement for the Popular Liberation of Angola
MPS	maritime pre-positioning squadron
MRB	motor rifle brigade
MRIK	Mission Role Installation Kit
MVD	Ministry of Internal Affairs
NATO	North Atlantic Treaty Organization
NBC	nuclear, biological, and chemical
NCA	National Command Authorities
NCO	noncommissioned officer
NGO	nongovernmental organization

NVA	North Vietnamese Army
OEF	Operation Enduring Freedom
OIF	Operation Iraqi Freedom
OP	observation post
OPLAN	operation plan
OPSEC	operations security
OSS	Office of Strategic Services
PDF	Panamanian Defense Force
PDPA	People's Democratic Party of Afghanistan
PSO	peace support operations
PUIC	Project Unique Identification Code
PzKpfw	*Panzerkampfwagen*
QRF	quick reaction force
RA	Republic of Afghanistan
RAAF	Royal Australian Air Force
RCT	regimental combat team
RPG	rocket-propelled grenade
RPV	remotely piloted vehicle
RSTA	reconnaissance, surveillance, and target acquisition
RWS	remote weapon station
SAA	South African Army
SAAF	South African Air Force
SAM	surface-to-air missile

SBCT	Stryker brigade combat team
SIGINT	signals intelligence
SSTOL	supershort-takeoff and -landing
SWA	Southwest Asia
SWAPO	Southwest African People's Organization
TNI	*Tentara Nasional Indonesia*
TO	table of organization
TOW	Tube-Launched, Optically Tracked, Wire-Guided
TRADOC	United States Army Training and Doctrine Command
TTP	tactics, techniques, and procedures
UHF	ultra high frequency
UN	United Nations
UNAMET	United Nations Mission in East Timor
UNITA	National Union for the Total Independence of Angola
UNITAF	Unified Task Force
UNOSOM	United Nations Operation in Somalia
UNPROFOR	United Nations Protection Force
UNTAET	United Nations Transitional Administration East Timor
USCINCCENT	Commander in Chief, United States Central Command
USSR	Union of Soviet Socialist Republics

VC	Viet Cong
VDV	*Vozdushno-Desantniy Voisk*
VHF	very high frequency
WMD	weapons of mass destruction

CHAPTER ONE

Introduction

In accordance with a late-1990s institutional decision to "shed [its] cold war designs in order to prepare . . . for the crises and wars of the 21st Century,"[1] the U.S. Army is in the midst of transforming itself to the Future Force. The term "cold war designs" is shorthand for the Army's recognition of the limitations, following the 1999 war in Kosovo, of its heavy and light forces in a security environment that it believes requires expeditionary—rather than forward-stationed—forces. The U.S. Army believes that heavy forces, although survivable and lethal, are slow to deploy and difficult to sustain. On the other hand, the U.S. Army's rapidly deployable light forces lack staying power, lethality, survivability, and tactical mobility.[2] Finally, the U.S. Army has possessed no air-droppable forced-entry armor capability since the retirement of the M551 Sheridan armored reconnaissance vehicle.

The Army Future Force

U.S. Army transformation has a clear purpose: "to produce a campaign-quality Army with joint and expeditionary capabilities, which will remain a vital and indispensable member of the Joint Force."[3] Due

1 U.S. Department of the Army, "The United States Army Vision," n.d.

2 Erik K. Shinseki, "Address to the Eisenhower Luncheon," 45th Annual Meeting of the Association of the United States Army, October 12, 1999.

3 U.S. Department of the Army, *2005 Army Modernization Plan*, Washington, D.C.: Headquarters, Department of the Army, 2005, p. 4.

to its experiences in Operation Enduring Freedom (OEF) in Afghanistan and Operation Iraqi Freedom (OIF) in Iraq, the U.S. Army has made significant progress toward the goals of coping with the realities it faces in today's operational environment and preparing for the future. The U.S. Army is well into a major restructuring effort in which it is "modularizing" from a division- to a brigade-based force. It is creating combined-arms brigade combat teams (BCTs) and other modular units that will increase unit readiness through a new U.S. Army Force Generation (ARFORGEN) process.[4] Thus, the U.S. Army will have "a larger pool of units to fulfill strategic commitments."[5] Adding National Guard brigades to the mix will further enhance U.S. Army force generation capabilities.

The U.S. Army is also shedding some of its "Cold War structure." These efforts involve

> decreasing the number of field artillery, air defense, engineer, armor and ordnance battalions while increasing military police, transportation, petroleum and water distribution, civil affairs, psychological operations and biological detection units.[6]

These measures will increase the service's capacity for stability and support operations.[7]

---

[4] See U.S. Department of the Army, *2007 Posture Statement*, Addendum H (Army Force Generation), February 14, 2007, which notes that ARFORGEN

> is the structured progression of increased unit readiness over time resulting in recurring periods of availability of trained, ready, and cohesive units. These units are prepared for operational deployment in support of Combatant Commanders' or civil authorities' requirements. Units are task organized in modular expeditionary forces, tailored for mission requirements. They are sustainable and have the capabilities and depth required to conduct the full range of operations in a persistent conflict. Operational requirements drive the ARFORGEN training and readiness process. ... The goal is to achieve a sustained, more predictable posture to generate trained and ready modular forces.

[5] U.S. Department of the Army, "Army Campaign Plan Briefing," n.d.

[6] Anne Plummer, "Army Chief Tells President Restructuring Force Could Cost $20 Billion," *Inside the Army*, February 9, 2004, p. 2.

[7] See U.S. Department of the Army, Field Manual (FM) 3-0: *Operations*, 2001, p. I–16. Stability and support capabilities are particularly important in conducting security, tran-

The centerpiece of U.S. Army transformation is the Future Combat Systems (FCS)–equipped combined-arms BCT:

> The FCS will comprise a key modular capability, with the strategic agility of light forces and the lethality, tactical mobility, and survivability of our heavy forces. FCS brigade combat teams will be the component of the modular Future Force most capable of implementing all aspects of [the U.S. Army's future] operational concept, particularly intratheater operational maneuver. The FCS further encompasses a set of technologies and capabilities that will spiral into the entire Army as they mature. Networked C4ISR [command, control, communications, computers, intelligence, surveillance, and reconnaissance], precision munitions, and advanced fire control will also be key enablers.[8]

The U.S. Army is also looking to the FCS to improve the "strategic responsiveness and deployability of the force overall as a result of its weight and cube advantages over current [heavy] systems."[9] Furthermore, the U.S. Army expects the FCS to provide an intratheater mobility capability. TRADOC [Training and Doctrine Command] Pamphlet 525-3-0, *The Army in Joint Operations: The Army's Future*

---

sition, and reconstruction operations and counterinsurgency. The U.S. Army's doctrine is designed to address the range of military operations though full spectrum operations, as stated below:

> When conducting full spectrum operations, commanders combine and sequence offensive, defensive, stability, and support operations to accomplish the mission. The JFC [joint force commander] and the Army component commander for a particular mission determine the emphasis Army forces place on each type of operation. Throughout the campaign, offensive, defensive, stability, and support missions occur simultaneously. As missions change from promoting peace to deterring war and from resolving conflict to war itself, the combinations of and transitions between these operations require skillful assessment, planning, preparation, and execution.

[8] U.S. Department of the Army, TRADOC Pamphlet 525-3-0: *The Army in Joint Operations: The Army's Future Force Capstone Concept, 2015–2024,* Version 2.0, Fort Monroe, Va.: Headquarters, U.S. Army Training and Doctrine Command, 2005, p. 38.

[9] U.S. Department of the Army, TRADOC Pamphlet 525-3-0: *The Army in Joint Operations*, p. 57.

Force Capstone Concept, 2015–2024, Version 2.0, specifically states a need for such a capability:

> Vertical maneuver of *mounted* forces, employing SSTOL [supershort-takeoff and landing] or HLVTOL [heavy-lift vertical-takeoff and landing] aircraft, puts large areas at risk for the adversary and will often lead to rapid tactical decision, shortening durations of battle, and contributing to the more rapid disintegration of the enemy force.[10]

Strategic deployability and, in particular, air transportability imperatives, clearly limit the weight of any potential FCS and make it a medium-armored force within the context of existing U.S. Army heavy and light forces. Pending the fielding of the Future Force, the U.S. Army is bridging the operational gap between light and heavy forces with the Stryker BCT (SBCT).[11] For SBCTs, the U.S. Army has chosen the C-130–transportable "Stryker" Light-Armored Vehicle (LAV)–III wheeled armored vehicle.[12]

Thus, the U.S. Army's transformation plans, both now with the SBCTs and in the future with the FCS BCT, are fundamentally linked to developing and fielding medium-armored forces. These medium-armored forces will have different characteristics and capabilities than the light and heavy units that the U.S. Army currently employs. Aside from the lessons being learned by the first SBCTs during their fielding at Fort Lewis, Washington, and in active combat operations in Iraq, there is little resident medium-armored force experience in the U.S. Army across the full range of military operations. Yet medium-

---

[10] U.S. Department of the Army, TRADOC Pamphlet 525-3-0: *The Army in Joint Operations*, p. 23.

[11] U.S. Department of the Army, "The United States Army Vision."

[12] Gary Sheftick and Michele Hammonds, "Army Selects GM to make Interim Armored Vehicles," Army News Service, November 20, 2000. See also John Gordon IV, David E. Johnson, and Peter A. Wilson, "Air-Mechanization: An Expensive and Fragile Concept," *Military Review*, Vol. 87, No. 1, January–February 2007, p. 69. The Stryker is not deployable by C-130s over long distances because of its weight; it must use improved airfields rather than field landing strips. Furthermore, the Stryker, when outfitted with add-on armor, will not fit on a C-130.

armored forces have been employed extensively in the 20th century, both by the United States and other nations, and across the range of military operations. Thus, the ultimate goal of this study is to provide insights from past medium armor operations—both positive and negative—to help inform U.S. Army decisionmaking about the Future Force and the FCS.

The Past as Prologue

Historically, three fundamental variables have affected the design of combat vehicles: lethality, survivability, and mobility. For centuries—from Assyrian charioteers to the mounted knights of the Middle Ages—improvements in lethality and mobility were fundamentally constrained by the requirement to rely on animal power for mobility and muscle-powered weapons for lethality. Horses pulled chariots and carried knights. The accuracy and power of weapons depended largely on the skill and strength of the charioteer, cavalryman, or knight with his individual weapons. Survivability was achieved through speed or through the protection of the warrior and his platform, be it chariot or horse, with armor.[13]

The invention of the internal combustion engine, coupled with new processes for producing steel (armor) and advances in armaments, set the stage for truly revolutionary changes in mounted combat. Machines began replacing muscle power.[14] By the first decade of the 20th century, machine-age combat vehicles began appearing in several European armies.[15] These early vehicles were wheeled armored cars that

---

[13] Kenneth Macksey and John H. Batchelor, *Tank: A History of the Armoured Fighting Vehicle*, New York: Charles Scribner's Sons, 1970, p. 5.

[14] Duncan Crow and Robert J. Icks, *Encyclopedia of Tanks*, Secaucus, N.J.: Chartwell, 1975, p. 9. Crow and Icks note of particular importance the introduction of the Paixhans gun with its explosive shell, the cast-steel rifled cannon, and the machine gun.

[15] See Bernard Brodie and Fawn Brodie, *From Crossbow to H-Bomb: The Evolution of the Weapons and Tactics of War*, rev. ed., Bloomington, Ind.: Indiana University Press, 1975, p. 196. The Brodies note that

generally weighed 4 tons or less and were armed with machine guns or small-caliber cannons.

Acceptance of these early armored cars by the militaries of the day was not, however, broad. Although the Italian Army employed armored cars in the Balkans and in the Tripolitanian desert, armored vehicles were largely confined to supporting roles, such as reconnaissance, mobile antiballoon guns, or machine-gun carriers.[16] Conservative European armies, wedded to offensive strategies enabled by mass armies of infantry, artillery, and cavalry, thought "proposals to develop petrol-driven vehicles beyond the requirements of transport into the realms of combat . . . [went] beyond the limits of sanity."[17] The early stalemate on the Western Front during World War I seemingly buttressed this aversion to machines on the battlefield. The limitations of wheeled armored cars on the World War I battlefield were insurmountable; these early vehicles simply could not negotiate the complex terrain of trenches, barbed wire, mud, and shell craters.[18]

The invention in 1915 of track-laying armored vehicles, called "tanks" as a deceptive measure by their British inventors, once again placed mounted soldiers in the thick of battle. The tank was designed to facilitate the advance of infantry on a stalemated battlefield dominated by machine guns and artillery. As with the knights of the agrarian age, mounted combat vehicles of the industrial age still faced the fundamental challenge of balancing the variables of lethality, survivability, and mobility. The first tank used in combat, the British Mark I, carried two 57-mm guns and four machine guns for lethality. Ten millimeters of armor provided survivability. Mobility was constrained

---

the notion of an armored car was at least as old as Leonardo [da Vinci]. The idea of making a machine gun mobile had been developed in 1898 by F. R. Simms, who successfully mounted a Maxim gun on a motorcycle. . . . Turreted armored cars built by the firm of Charron, Giradot and Voight in France had been sold to the Russians as early as 1904.

[16] Macksey and Batchelor, *Tank: A History of the Armored Fighting Vehicle*, p. 10; Crow and Icks, *Encyclopedia of Tanks*, p. 10.

[17] Macksey and Batchelor, *Tank: A History of the Armored Fighting Vehicle*, p. 10.

[18] Macksey and Batchelor, *Tank: A History of the Armored Fighting Vehicle*, p. 13.

by the tank's weight of 28 tons and its small 105-HP Daimler engine, whose maximum speed was 3.7 miles per hour and range of operation was limited to 23 miles.[19]

The first use of tanks in combat occurred on September 15, 1916, at the Battle of Flers-Courcelette. Forty-nine British Mark I tanks assembled as part of British General Sir Douglas Haig's renewed Somme Offensive. Because of mechanical difficulties, only 32 of the crude tanks moved forward with the attacking infantry and only nine of these covered the distance to the German lines. Nevertheless, the British tanks caused panic in the German ranks and their impressive performance caused General Haig to establish a separate headquarters for the new weapon.[20]

The British followed this initial use of tanks with a much larger effort at Cambrai on November 20, 1917. On that day, 376 Mark IV tanks broke through German lines and penetrated 4 miles into the enemy's defenses—gains unprecedented on the Western Front since the initial stalemate in 1914. Unfortunately for the British cause, the horse cavalry exploitation force was stymied by its vulnerability to remaining German machine guns.[21]

Cambrai also witnessed the emergence of antitank doctrine and weapons. By the end of the war, improvements in tank obstacles, antitank guns, and antitank tactics rendered the slow, mechanically unreliable, and relatively lightly armored tanks of the day extremely

---

[19] Macksey and Batchelor, *Tank: A History of the Armored Fighting Vehicle*, p. 25; Brodie and Brodie, *From Crossbow to H-Bomb*, p. 197.

[20] David E. Johnson, *Fast Tanks and Heavy Bombers: Innovation in the U.S. Army, 1917–1954*, Ithaca: Cornell University Press, 1998, pp. 30–31. On this first use of tanks in combat see also Tim Travers, *The Killing Ground: The British Army, the Western Front, and the Emergence of Modern Warfare, 1900–1918*, London: Allen and Unwin, 1987, pp. 166–167, 179–181; and John Keegan, *The Illustrated Face of Battle*, New York: Viking, 1988, p. 183. On the development of early armored vehicles, see Macksey and Batchelor, *Tank: A History of the Armored Fighting Vehicle*; Crow and Icks, *Encyclopedia of Tanks*; Chris Ellis and Peter Chamberlain, *The Great Tanks*, London: Hamlyn Publishing Group, 1975; Kenneth Macksey, *Tank Versus Tank: The Illustrated Story of Armored Battlefield Conflict in the Twentieth Century*, Topsfield, Mass.: Salem House, 1988; and A. J. Smithers, *A New Excalibur: The Development of the Tank, 1909–1939*, London: L. Cooper in association with Secker & Warburg, 1986.

[21] Johnson, *Fast Tanks and Heavy Bombers*, p. 30.

vulnerable, unless accompanying infantry or fires dealt with antitank defenses.[22]

In the aftermath of World War I and to this day, nations have grappled with the military implications of combat vehicles and the issues of defending against them. Function has driven form, and combat vehicles used for close combat generally have been designed (or called upon) to perform one or more of the following roles:

- Serve as the basis of mobile armored operations or direct-fire platforms (i.e., tanks).
- Support infantry operations (e.g., mobile assault guns, infantry armored personnel carriers, infantry fighting vehicles [IFVs]).
- Perform traditional cavalry missions (e.g., reconnaissance, screening, raiding, exploitation, pursuit).
- Provide mobile antitank platforms (cannon or missile).

In developing vehicles to perform these functions, designers have had to make trade-offs between the variables of lethality, survivability, and mobility—principally because of the issue of weight. Weight has been a limiting factor because it affects deployability, trafficability, and vehicle speed. Quite simply, adding armor protection for enhanced survivability increases system weight, as does incorporating larger weapons for increased lethality. As seen in the cases analyzed in this study, witting and unwitting trade-offs made by armies over the decades have affected the performance of fielded forces.

Monograph Objective and Parameters

This monograph is a qualitative assessment of the historical employment of medium-armored forces in the 20th century. We assess U.S. and foreign experiences to analyze how medium-armored forces performed in the past at several points along the range of military opera-

---

[22] Jonathan M. House, *Combined Arms Warfare in the Twentieth Century,* Lawrence, Kan.: University of Kansas Press, 2001, pp. 45–49.

tions, in complex terrain, and against different types of opponents. Our objective is to provide insights from the past employment of medium-armored forces to assist the U.S. Army in its efforts to develop operational concepts for future U.S. Army medium-armored forces and to inform choices about the technical characteristics of the FCS.

Tasks

The project's sponsor specified three central questions for the study:

1. *What unique capabilities have medium-armored forces brought to past conflicts, and where along the spectrum of operations have they been most valuable?* Tasks: Analyze 20th century cases of the employment of medium-armored forces across the spectrum of operations by different nations to provide a qualitative assessment of their performance. Determine measures of success, detect force shortcomings, and identify any measures taken by these forces to compensate for identified shortcomings.
2. *How have medium-armored forces performed in complex terrain in the past?* Tasks: Analyze U.S. and foreign experiences with medium-armored forces in complex (e.g., urban, jungle, mountainous, undeveloped infrastructure) terrain to assess their performance.
3. *What advantages has the rapid-deployment capability of medium-armored forces provided to operational commanders in the past?* Tasks: Assess historical examples of the rapid deployment of medium-armored forces in past conflicts to determine whether their early presence provided the operational commander with capabilities that, in the absence of a medium-armored force, would not have been available.

Definitions

What is a medium-armored force? During early research it became apparent that defining what constitutes a medium-armored force was a crucial element in deciding which cases to assess. The two definitions broadly used in this monograph are tied to platforms (e.g., tanks and other armored vehicles) and are also contextual. That is, a force is medium-armored

- *In the context of a nation's overall force.* For example, the U.S. Army's SBCT, composed of medium-weight Stryker wheeled armored vehicles, is medium-armored when compared to U.S. light infantry BCTs or heavy BCTs (with their Abrams tanks and Bradley fighting vehicles).[23]
- *In the context of the opponent's armored vehicles.* For example, the U.S. Army's M4 Sherman main battle tanks (MBTs) (the heaviest tank fielded by the U.S. Army for most of the war) in World War II were medium-armored relative to German heavy tanks. This was mainly due to lethality and survivability issues, which were manifested in weight, because increases in either resulted in more system weight given the technologies of the day.

Methodology

We used historical research, mainly as supplied in secondary sources, to select and develop the cases studied in this report. To provide a comprehensive survey of past use of medium armored vehicles, we included as many cases as possible. At the sponsor's request, we assessed each case from several perspectives:

- How were medium-armored forces employed, and why does this monograph define them as "medium-armored"?

---

[23] One of the principle reasons for fielding the Stryker and the FCS is the U.S. Army's desire to make its forces more deployable. In the case of the FCS, the U.S. Army believes it can—through the use of advanced technologies to improve situational awareness, survivability, and lethality—provide equivalent protection and lethality to existing heavy forces equipped with Abrams tanks and Bradley fighting vehicles, but with less weight. That said, one could argue that even more survivability and lethality could be added with more weight, using, for example, FCS technologies to upgrade existing tanks and fighting vehicles or by raising weight constraints for the FCS itself. Indeed, over the life of the FCS program, the allowable weight of the vehicles has continually increased to accommodate capabilities the Army requires in the FCS. See Stew Magnuson, "Future Combat Vehicles Will Fall Short of Preferred Weight," *National Defense*, Vol. XCI, No. 643, June 2007, pp. 16–17.

- What doctrine, organization, training, leadership and educataion, personnel, and facilities (DOTMLPF) insights emerge? This discussion is particularly important given the centrality of DOTMLPF to U.S. Department of Defense (DoD) joint capabilities development.[24] Instead of assessing the facilities aspects of each case separately, the study covers in-theater implications in a discussion of the combat service support (CSS) system of the battlefield operating system (BOS). (The joint definitions of DOTMLPF are provided in Appendix C.)
- What are the BOS (maneuver, fire support, air defense, command and control [C2], intelligence, mobility/countermobility/survivability, and CSS) implications? (The U.S. Army BOS definitions used in this study are provided in Appendix C.)[25]
- Which operational characteristics that the U.S. Army expects of a transformed force (i.e., responsiveness, deployability, agility, versatility, lethality, survivability, and sustainability) surfaced, or did not? (The U.S. Army definition of a transformed force is provided in Appendix C.)
- How did the medium-armored force under examination perform in the case environments (i.e., terrain)?
- What key insights emerge?

Finally, we describe overarching insights that are common among cases.

---

[24] We use DOTMLPF definitions from Chairman of the Joint Chiefs of Staff, Instruction 3170.01E, "Joint Capabilities Integration and Development System," May 11, 2005, pp. GL-9 to GL-10. This instruction was replaced by Chairman of the Joint Chiefs of Staff, Instruction 3170.01F, "Joint Capabilities and Integration Development System," May 1, 2007, which does not include DOTMLPF definitions.

[25] This monograph uses the BOS definitions that were current at the time the study began. The authors recognize this construct has changed.

Table 1.1
Case Studies

Case	Complex Terrain	Point in the Range of Military Operations	Type of Operation	Types of Armored Vehicles and Other Forces
Armored warfare during the Spanish Civil War (1936–1939)	Broken and mountainous; urban	High	Major operations (civil war with external support to both sides)	German and Italian medium-armored vs. Soviet Union heavy
U.S. armored divisions in France and Germany during World War II (1944–1945)	Urban; hedgerows; forests	High	Major operations	U.S. medium-armored vs. German heavy
Armored cavalry and mechanized infantry in Vietnam (1965–1972)	Jungle	High	Major operations; counterinsurgency operations	U.S. medium-armored and heavy vs. Viet Cong and North Vietnamese light
Soviet airborne operations in Prague, Czechoslovakia (1968)	Urban	Middle	Strike (regime change)	Soviet Union medium-armored and heavy vs. Czechoslovakian light (mainly civilian forces)
South Africa in Angola (1975–1988)	Close; undeveloped infrastructure	Middle	Major operations; raids	South African medium-armored vs. Angolan heavy
Soviet Union in Afghanistan (1979–1989)	Urban; mountains; undeveloped infrastructure	Middle	Strike (regime change); counterinsurgency operations	Soviet Union medium-armored and heavy vs. Afghan light
Operation Just Cause, Panama (1989)	Urban	Middle	Strike (regime change)	U.S. medium-armored vs. Panamanian medium and light

Table 1.1—Continued

Case	Complex Terrain	Point in the Range of Military Operations	Type of Operation	Types of Armored Vehicles and Other Forces
1st Marine Division light armored vehicles (LAVs) in Operations Desert Shield and Desert Storm, Southwest Asia (1990–1991)	Desert; limited visibility	High	Major operations	U.S. (Marine Corps) medium-armored vs. Iraqi heavy and medium
Task Force Ranger in Mogadishu, Somalia (1993)	Urban	Low	Raid	U.S. light and coalition (Malaysian and Pakistani) medium-armored vs. Somali light
Russia in Chechnya I (1994–1996) and II (1999–2001)	Urban; mountains	Middle	Counterinsurgency operations; combating terrorism	Russian medium-armored and heavy vs. Chechen light
Australia and New Zealand in East Timor (1999–2000)	Urban; jungle; undeveloped infrastructure	Low	Peace operations	Australian and New Zealand medium-armored vs. rebel light
SBCTs in Operation Iraqi Freedom (OIF) (2003–2005)	Urban	Middle	Counterinsurgency operations; combating terrorism	U.S. medium-armored vs. indigenous Iraqi and foreign fighter light

Cases

The cases selected for this study, shown in Table 1.1, facilitated answering the sponsor's three questions.[26]

Again, each case was viewed through three prisms: DOTMLPF, BOS, and the characteristics that the U.S. Army desired of its transformed forces according to its 1999 statement that it wanted in its transformed forces. Finally, the cases represent the historical use of medium-armored forces across the range of military operations shown in Figure 1.1.[27]

Monograph Organization

Historical case assessments of the use of medium-armored forces are discussed in three chapters that are organized along the range of military operations depicted in Figure 1.1. These three chapters and broad types of military operation are (1) operations at the high end of the range, (2) operations in the middle of the range, and (3) operations at the lower end of the range. Within each of the three chapters, the cases are presented chronologically. The monograph's final chapter offers insights for the U.S. Army Future Force based on the case analysis. Three appendixes provide additional analysis and supporting information. Appendix A synthesizes medium-armored force insights related to DOTMLPF, BOS, characteristics of a transformed force, and performance in complex terrain. Appendix B contains the detailed case assessments. Appendix C provides definitions.

---

[26] Three other cases were briefly examined (India vs. Pakistan, 1965; Libya vs. Chad, 1981; and the Falklands War, 1982) but ultimately discarded because medium forces did not play a significant role. Additionally, the case studies presented in this monograph do not discuss medium forces that were used mainly for reconnaissance or indirect fire support. We focused on cases where medium forces served as ground maneuver elements.

[27] Note that Table 1.1 also shows the type(s) of military operations conducted by medium-weight armored forces in each of the cases, taken form the following list in U.S. Joint Chiefs of Staff, Joint Publication 3-0, p. I-7: major operations; homeland defense; civil support; strikes; raids; show of force; enforcement of sanctions; protection of shipping; freedom of navigation; peace operations; support to insurgency; counterinsurgency operations; combating terrorism; noncombatant evacuation operations; recovery operations; consequence management; foreign humanitarian assistance; nation assistance; arms control and disarmament; and routine, recurring military activities.

Figure 1.1
The Range of Military Operations

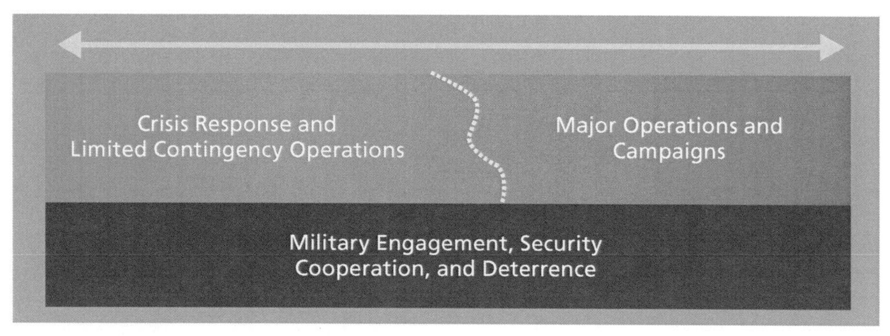

SOURCE: Adapted from U.S. Joint Chiefs of Staff, Joint Publication 3-0, *Joint Operations*, 2006, p. I-8.
RAND *MG709-1.1*

CHAPTER TWO

Medium-Armored Forces in Operations at the High End of the Range of Military Operations

This chapter examines four instances of the use of medium-armored forces in operations at the high end of the range:

- the Spanish Civil War (1936–1939), a limited conventional conflict
- U.S. armored divisions in France and Germany during World War II (June 1944–May 1945), a global conventional conflict
- the United States in Vietnam (1965–1972), a limited conventional conflict and counterinsurgency with major operations and campaigns against regular North Vietnamese Army (NVA) forces and insurgents
- the 1st Marine Division's LAV during Operations Desert Shield and Desert Storm (1990–1991), operations that constituted a major theater war.

Armored Warfare in the Spanish Civil War (1936–1939)

In July 1936 civil war broke out in Spain. Nationalist forces supported by Italy and Germany fought numerically superior Republican forces who were aided by France, the Soviet Union, and international volunteer units. The war ended in January 1939 with a Nationalist victory.

The Spanish Civil War was the first conflict in which the opposing sides each possessed armored vehicles in significant numbers. The war also witnessed the first tank-versus-tank battle in March 1937 at Guadalajara. This case offers insights into early armored vehicle design and doctrine and an opportunity to assess the performance of medium-armored Nationalist forces (Italian and German) against heavy-armored Republican (Soviet) forces.

The Armored Forces

The principal armored vehicles available to the Nationalist forces were the German *Panzerkampfwagen* (PzKpfw) I tank and the Italian L3/35 tankette. The Republican forces employed several Soviet armored vehicles: T-26 and BT-5 tanks and the BA-10 armored car. Approximately 180 German tanks, 150 Italian tankettes, and 331 Soviet tanks were shipped to Spain during the Spanish Civil War.[1] Table 2.1 shows the characteristics of these vehicles. Although all were relatively lightly armored and thus vulnerable, the Soviet tanks enjoyed the significant advantage of more-lethal 45-mm or 37-mm guns as compared to the machine guns that served as the primary armament on the German and Italian vehicles.[2] This lethality advantage was not without cost, however: Soviet tanks weighed significantly more than their German and Italian opponents. From a comparative perspective, Soviet tanks were heavy and German and Italian tanks were medium-armored.

---

[1] Hugh Thomas, *The Spanish Civil War*, New York: Harper and Row, 1961, pp. 634–43; John L. S. Daley, "The Theory and Practice of Armored Warfare in Spain: October 1936–February 1937," *Armor*, Vol. 108, No. 2, March–April 1999, p. 43; Steven J. Zaloga, "Soviet Tank Operations in the Spanish Civil War," *The Journal of Slavic Military Studies*, Vol. 12, No. 3, September 1999, pp. 134–162. Both Thomas and Daley estimate that the Soviet Union supplied approximately 700 tanks. Zaloga's figure of 331 is based on research in Soviet archival material. See also Wendell G. Johnson, "The Employment of Supporting Arms in the Spanish Civil War," *C. & G.S.S. Quarterly*, Vol. 19, No. 72, March 1939, p. 13. Johnson notes that "The Spanish Army on 17 July 1936, had practically no modern tanks and only 120 obsolete or obsolescent machines, most of which were war-time [World War I] Renaults. Crude armored cars were built on truck chassis."

[2] See also S. Hart and R. Hart, *German Tanks of World War II*, New York: Barnes and Noble, 1999, p. 14. This work notes that "Spanish workshops modified a few Panzer I tanks to mount a more potent 20mm cannon, although this upgrading resulted in a markedly reduced cross-country performance."

Table 2.1
Armored Vehicles in the Spanish Civil War

System	Weight (tons)	Armament	Max Armor (mm)
German PzKpfw-I tank	5.5	Two 7.92-mm machine guns	15
Italian L3/35 tankette	3.6	Two 8-mm machine guns	12
Soviet T-26 tank	9.0	45-mm gun; two 7.62-mm machine guns	15
Soviet BT-5 tank	11.2	45-mm gun; 7.62-mm machine gun	13
Soviet BA-10 armored car	5.2	37-mm gun; 7.62-mm machine gun	15

Employment

Germany, the Soviet Union, and Italy all deployed partially manned armored formations to Spain, expecting that Spanish soldiers would fill out the formations. The German forces were in the *Imker Drohne* group, the Soviets in the Krivoshein Detachment (named for its commander, Semyon M. Krivoshein), and the Italians in the *Corpo Truppe Volontarie*, or Corps of Voluntary Troops.[3]

Throughout the war, tank units were generally employed in an infantry support role in battalion-, company-, or smaller-sized units. The limited number of tanks available largely dictated unit size. Furthermore, by the time the majority of the foreign tank forces arrived in Spain, the strategic situation was generally stalemated. As one observer noted,

> The tendency, until one side was exhausted, would always be toward stabilization; fronts would be broken through and rebuilt

---

[3] Daley, "The Theory and Practice of Armored Warfare in Spain," p. 30; Brian R. Sullivan, "Fascist Italy's Military Involvement in the Spanish Civil War," *The Journal of Military History*, Vol. 59, No. 4, October 1995, p. 706. See also Brian Sullivan, "The Italian Armed Forces, 1918–1940," in Allan R. Millet and Williamson Murray, eds., *The Interwar Period*, Boston: Unwin Hyman, 1990.

> repeatedly; and the possibility of a reversion to an outright war of position would always be strong.[4]

Thus, the war became a series of Republican and Nationalist offensives to capture key cities, particularly Madrid on the part of the Nationalists. Tanks functioned in supporting roles during combat operations, not in the independent, decisive roles envisioned by some. They were employed with some level of effectiveness, generally in an infantry support role, throughout the war.

During the Nationalist March 1937 Guadalajara offensive against Madrid, the first tank versus tank battle occurred. Italian turretless tankettes, armed with fixed, forward-firing machine guns, were decimated by Soviet T-26 tanks armed with 45-mm cannons.[5] Similarly, the German PzKpfw I, also armed only with machine guns, proved no match for Soviet tanks or for the Soviet BA-10 armored cars, armed with 37-mm cannons. Indeed, after several confrontations with Soviet tanks and armored cars, the German commander of the *Imker Drohne*, Wilhelm Ritter von Thoma, directed his "personnel to avoid engagements with Soviet tanks whenever possible."[6]

Each nation that provided armored forces to the Spanish Republicans and Nationalists produced views about how to employ large, massed, armored formations in rapid, decisive operations. The Germans were developing the blitzkrieg.[7] The Soviets had embraced concepts stressing "decisive victory . . . by offensive action in depth," and those concepts were codified in the Provisional Field Service Regulations of 1936.[8] The Italians became committed to a theory of *guerra di*

---

[4] Earl F. Ziemke, "The Soviet Armed Forces in the Interwar Period," in Millet and Murray, eds., *The Interwar Period*, p. 30.

[5] Johnson, "The Employment of Supporting Arms," pp. 7–8, 13. See also Antony Beevor, *The Battle for Spain: The Spanish Civil War, 1936–1939*, New York: Penguin Books, 2006, p. 427. Beevor notes that the Italian tanks "looked and performed more like a clockwork toy."

[6] John L.S. Daley, "Soviet and German Advisors Put Doctrine to the Test: Tanks in the Siege of Madrid," *Armor*, Vol. 108, No. 3, May–June 1999, p. 35.

[7] Daley, "The Theory and Practice of Armored Warfare in Spain," pp. 39–40.

[8] Daley, "The Theory and Practice of Armored Warfare in Spain," p. 40.

rapido corso.[9] The stabilized conditions that existed when foreign formations intervened in the war, coupled with the relatively small numbers of armored vehicles deployed, created circumstances where these various theories of rapid, decisive operations could not be executed. Instead, armored vehicles became tactical weapons normally employed in support of limited offensive operations or to bolster defenses.

What did develop over time in Spain was an appreciation among all the forces involved of the importance of the contributions of various arms in offensive combat. This was reported by a U.S. officer in 1939:

> In most offensive operations carried out by either side during the past year or more, tanks seem to have been the third echelon of the attack. Aviation and artillery strike the first and second blows, tanks the third, infantry the fourth, and cavalry enters the action as the fifth and final echelon to pursue, outflank, or mop-up.[10]

The same officer noted, however, that tanks had shown some value in pursuit and as a counterattack force "if used before the enemy has organized the newly won terrain and brought forward his antitank weapons."[11] Tanks also participated in operations in villages and cities. On the offensive, given their light armor, they were "most vulnerable to grenades and often makeshift antitank measures."[12] Tanks were useful, however, in a "fire brigade" role in cities. This was particularly apparent in the extended Republican defense of Madrid, where Soviet T-26s "were mobile enough to appear at any threatened point and well enough armed to make a crucial difference once there."[13]

Nevertheless, one lesson was very clear: Armored forces, even during limited breakthroughs and exploitations achieved during the

---

[9] Sullivan, "The Italian Armed Forces," in Millet and Murray, eds., *The Interwar Period*, p. 706.

[10] Johnson, "Employment of Supporting Arms," p. 13.

[11] Johnson, "Employment of Supporting Arms," p. 13.

[12] Daley, "Soviet and German Advisors Put Doctrine to the Test," p. 34.

[13] Daley, "Soviet and German Advisors Put Doctrine to the Test," p. 36.

Spanish Civil War, required competent infantry support to negate antitank defenses.[14] Thus, combat in Spain showed that

> whatever promise independent mechanized action held at the operational and strategic levels, frequent combined-arms operations involving tanks and dismounted infantry were to be expected.[15]

Antony Beevor believes that the Germans learned something at the higher level of operations:

> Their tanks needed to be more heavily armed and concentrated in armoured divisions for '*Schwerpunkt*' breakthroughs. . . . [I]t was as a result of the war in Spain that the German army saw the need to increase the size and power of its tank force.[16]

The effect of the war on Soviet concepts for armored warfare was much more constrained:

> The purging of Marshal [Mikhail Nikolayevich] Tukhachevsky and his supporters who advocated the new approach to armored warfare returned communist military theory to the political safety of obsolete tactics.[17]

---

[14] Daley, "Soviet and German Advisors Put Doctrine to the Test," p. 36.

[15] Daley, "The Theory and Practice of Armored Warfare in Spain," p. 42. See also Antonio J. Candil, "Soviet Armor in Spain: Aid Mission to Republicans Tested Doctrine and Equipment," *Armor*, Vol. 108, No. 2, March–April 1999, p. 38. The author concluded in this article that the Spanish Civil War experience showed that "[t]anks needed to be supported by motorized infantry. Failing to do that caused many of the Soviet mistakes. Only in rare cases, or against limited objectives, should tanks proceed alone." Furthermore, "[a] great advantage accrued to close cooperation with aircraft, which could aid command and control, provide combat support, and perform reconnaissance."

[16] Beevor, *The Battle for Spain*, p. 427.

[17] Beevor, *The Battle for Spain*, p. 196.

Thus, even during the war in Spain, "Soviet advisers could not advocate modern armoured tactics after the show trial of Marshal Tukhachevsky."[18]

The Spanish Civil War offered several lessons for the design of armored vehicles. The lightly armored tanks employed in the conflict were vulnerable to other tanks and to antitank weapons, ranging from antitank guns to field expedient devices (such as what would become known as Molotov cocktails). The conditions in Spain of "battle-torn terrain" and "natural and artificial obstacles" made the going very difficult for tanks and made speed "unusable."[19] Thus, the need for protection militated for armor over speed: "Most foreign commentators now stress armor above speed. Certainly, if one or the other has to be sacrificed, speed must give way to armor."[20] Additionally, it became apparent that tanks needed turrets that could traverse 360 degrees (to address flank threats) and guns and accurate fire-control systems capable of dealing with other tanks. On-board tank radios were necessary to command and control tank units.[21] Finally, the superiority of the Soviet T-26 and BT-5 in the Spanish Civil War derived from the greater lethality at longer ranges of their main guns when compared to the German PzKpfw I and Italian tankettes, a lesson that would also affect tank design in the coming Second World War.

Key Insights

Although the Spanish Civil war did not afford the most ardent of the interwar theorists of armor with an environment in which to test their theories of large-scale armored warfare, it did yield insights about the design of armored vehicles and concepts for their use:

- Armed with weapons ranging from machine guns to cannons, armored vehicles provided protected, mobile firepower and provided useful support to light infantry. Furthermore, as in the

---

18 Beevor, *The Battle for Spain*, p. 427.

19 Johnson, "Employment of Supporting Arms," pp. 13–14.

20 Johnson, "Employment of Supporting Arms," p. 16.

21 Daley, "The Theory and Practice of Armored Warfare in Spain," pp. 41–42.

Republican defense of Madrid, armored vehicles provided a way to rapidly respond to changes in the tactical situation with mobile, protected firepower—a role light infantry could not perform.

- Armored vehicles required heavier armor to increase survivability, and vehicle speed was sacrificed to this end.
- Tanks required cannons capable of defeating other tanks, and required a turret that gave 360-degree coverage.
- Tank reliability and off-road capability needed improvement.
- Antimechanized defenses, both active and passive, posed a significant threat to armored vehicles.
- Tanks were most effective when employed with other arms.
- Radio communications were required to coordinate armored vehicle movement and integrate other arms.

U.S. Armored Forces Versus German Armored Forces in Western Europe During World War II (1944–1945)

The final battle for western Europe began on June 6, 1944, with the invasion of Normandy, and ended on May 8, 1945, with the unconditional surrender of the Nazi Third Reich. World War II witnessed the first large-scale use of armored forces in the type of mobile, combined-arms warfare that had been theorized about during the interwar period and modestly evaluated in the Spanish Civil War. Most famously, the German blitzkrieg offensives in Poland, France, and Russia came to epitomize this new style of warfare.

This case examines how medium-armored U.S. forces performed against heavy-armored German forces. It provides insights about the influence of U.S. Army interwar decisions about doctrine and combat vehicle design on combat performance during World War II.

The Armored Forces

U.S. Army Tanks and Tank Destroyers. The M4 Sherman medium tank, in several configurations, was the MBT available to U.S. armored formations. U.S. Shermans mounted a 75-mm gun or a more-powerful 76-mm gun, and most tank battalions in U.S. armored divisions had a mix of the two versions. In February 1945 the U.S. Army began fielding the M26 medium tank with a 90-mm gun, but these tanks made little operational contribution because only some 200 had been issued to troops when the war ended.[22]

U.S. medium tank design focused on providing a weapon "to destroy enemy personnel and automatic weapons."[23] Fighting enemy tanks was not their role. U.S. doctrine dictated that tank destroyer units would provide U.S. Army forces with "anti-mechanized protection," if required. To this end, the U.S. Army fielded tank destroyer units that were equipped with weapons ranging from 37-mm towed guns to 90-mm self-propelled guns. The self-propelled tank destroyers were lightly armored; this and their open turrets made them vulnerable to direct- and indirect-fire weapons (e.g., artillery). Table 2.2 shows the characteristics of World War II U.S. tanks and tank destroyers.

German Army Tanks and Tank Destroyers in 1944. German armored forces contained both medium and heavy tanks and tank destroyers. German armored vehicle technology evolved throughout World War II, and by the summer of 1944, the German Army had fielded the vehicles in Table 2.3. German armored vehicle design in 1944 reflected the realization that tanks had to fight other tanks and survive. The German Army learned this lesson in 1941, when its tanks could not cope with the Russian T-34 tank. The *Panzer* V Panther was the result of this learning experience. This 45-ton tank featured sloped frontal armor and thicker armor than its predecessors, both of which provided survivability, and its high-velocity 75-mm gun provided

---

22 Johnson, *Fast Tanks and Heavy Bombers*, pp. 189–201.

23 U.S. War Department, FM 17-100, *(Tentative) Employment of the Armored Division and Separate Units*, Washington, D.C., 1943, pp. 1, 8, 13. See also Christopher R. Gabel, "World War II Armor Operations in Europe," in George F. Hofmann and Donn A. Starry, eds., *From Camp Colt to Desert Storm: The History of U.S. Armored Forces*, Lexington, Ky.: University of Kentucky Press, 1999, p. 156. The targets envisaged for the Sherman tank are evident in the ammunition it carried:

> Ammunition stowage totaled ninety-seven rounds of 75mm ammunition, of which approximately 70 percent would typically be high explosive, 20 percent armor piercing, and 10 percent white phosphorous. The latter served as both an incendiary and a smoke munition.

Table 2.2
Principal U.S. Armored Vehicles

System	Weight (tons)	Armament	Max Armor (mm)	Penetration (mm of armor at yards)			
				500	1,000	1,500	2,000
M3 (light tank)	14.00	37-mm gun; three .30-caliber machine guns	51	53	46	40	35
M4 Sherman tank	33.25 to 35.50	75-mm gun; .50-caliber machine gun; two .30-caliber machine guns	81	75	69	55	40
M4 Sherman tank	35.00 to 36.50	76-mm gun; .50-caliber machine gun; two .30-caliber machine guns	81	158	134	117	99
M26 Pershing tank	46.00	90-mm gun; .50-caliber machine gun; .30-caliber machine gun	90	221	195	177	154
M10 tank destroyer	33.00	3-in. gun; .50-caliber machine gun; .30-caliber machine gun	59[a]	157	135	116	98
M18 tank destroyer	20.00	76-mm gun; .50-caliber machine gun; .30-caliber machine gun	25[a]	158	134	117	99
M36 tank destroyer	31.00	90-mm gun; .50-caliber machine gun; .30-caliber machine gun	51[a]	221	195	177	154

SOURCES: Lida Mayo, *The Ordnance Department: On Beachhead and Battlefront*, Washington D.C.: Office of the Chief of Military History, U.S. Army, 1968, pp. 334–336; Constance McLaughlin Green, Harry C. Thomson, and Peter C. Roots, *The Ordnance Department: Planning Munitions for War*, Washington, D.C.: Office of the Chief of Military History, Department of the Army, [1955] 1990, pp. 372–373; R. P. Hunnicutt, *Stuart: A History of the American Light Tank*, Vol. 1, Novato, Calif.: Presidio Press, 1992, pp. 480, 495–496; R. P. Hunnicutt, *Sherman: A History of the American Medium Tank*, Novato, Calif.: Presidio Press, 1994, pp. 562–567; Charles M. Baily, *Faint Praise: American Tanks and Tank Destroyers During World War II*, Hamden, Conn.: Archon Books, 1983, pp. 110, 147–156; Konrad F. Schreier, Jr., *Standard Guide to U.S. World War II Tanks and Artillery*, Iola, Wisc.: Krause Publications, 1994, pp. 15–33, 46–56; Roman Johann Jarymowycz, *Tank Tactics: From Normandy to Lorraine*, Boulder, Colo.: Lynne Rienner, 2001, p. 277.

[a] Open turret top.

Table 2.3
Principal German Armored Vehicles

System	Weight (tons)	Armament	Max Armor (mm)	Penetration (mm of armor at yards)			
				500	1,000	1,500	2,000
Panzer V Panther (Ausf G Sd Kfz 171)	44.80	75-mm gun; three 7.92-mm machine guns	120	174	150	127	106
Panzer VI Tiger I (Ausf H/E Sd Kfz 181)	56.90	88-mm gun; two 7.92-mm machine guns	100	155	138	122	110
Panzer VI Tiger II (Ausf B Sd Kfz 182)	69.40	88-mm gun; three 7.92-mm machine guns	185	217	193	170	152
Jagdpanzer "Marder III" (Ausf M Sd Kfz 138) tank destroyer	10.50	75-mm gun; 7.92-mm machine gun	20[a]	120	97	77	64
Jagdpanzer "Hetzer" (Pz 38[t]) tank destroyer	18.00	75-mm gun; 7.92-mm machine gun	75	120	97	77	64
StuG III (Ausf G Sd Kfz 142) assault gun and tank destroyer	22.00	75-mm gun; 7.92-mm machine gun	90	120	97	77	64
Jagdpanzer PzIV (Sd Kfz 162/1)	25.80	75-mm gun; 7.92-mm machine gun	90	174	150	127	106
Jagdpanther (Jgd Pz V Sd Kfz 173) tank destroyer	46.00	88-mm gun; 7.92-mm machine gun	90	217	193	171	153
Jagdtiger (Ausf B Sd Kfz 186)	75.68 to 77.66	128-mm gun; two 7.92-mm machine guns	250	178	167	157	148

SOURCES: Thomas L. Jentz, ed., *Panzertruppen: The Complete Guide to the Creation and Employment of Germany's Tank Force, 1933–1942*, Atglen, Pa.: Schiffer, 1996, pp. 292–296; Hart and Hart, *German Tanks of World War II*, pp. 70–81, 94–109, 116–43; Peter Chamberlain and Hilary Doyle, *Encyclopedia of German Tanks of World War Two: A Complete Illustrated Directory of German Battle Tanks, Armoured Cars, Self-Propelled Guns and Semi-Tracked Vehicles, 1933–1945*, London: Arms and Armour Press, [1978] 2001, pp. 58–145.

[a] Open turret top.

stand-off lethality. As Table 2.3 shows, all German tanks and tank destroyers followed this trend toward greater protection and lethality. All German tank destroyers, except the *Marder*, were heavily armored, included overhead protection, and mounted increasingly lethal guns as they evolved.

Employment

The landings on D-Day in Normandy marked the beginning of the endgame for World War II in the west. The war in western Europe was waged against a Nazi state that was "on the way to defeat thanks to Allied victories in the North Atlantic, in the skies over Europe, and especially on the eastern front."[24] Nevertheless, despite being largely on the defensive from D-Day to the end of the war, the German Army proved to be a formidable opponent. For the U.S. Army in Europe, the war had fairly distinct phases:

- landing on the continent (Normandy and southern France) and breaking out (June–July 1944)
- exploiting and pursuing the German army following the break-outs (July–September 1944)
- fighting the battle of attrition after the exploitation and pursuit stalled west of the Rhine River (September–December 1944)
- reacting to the German counteroffensives in the Ardennes and the Vosges and closing to the Rhine River (December 1944–March 1945)
- fighting the final offensive against Germany after establishing beachheads across the Rhine River (March 1945–May 1945).[25]

---

[24] Gabel, "World War II Armor Operations in Europe," in Hofmann and Starry, eds., *From Camp Colt to Desert Storm*, pp. 160.

[25] The literature on World War II is vast, but the series *The U.S. Army in World War II*—also known as "The Green Books"—published by the U.S. Army Center of Military History is still a superb reference. Its volumes cover in considerable detail the breadth of the U.S. Army's experience across what is now called the DOTMLPF. A superb discussion of the last year of World War II can be found in Max Hastings, *Armageddon: The Battle for Germany, 1944–1945*, New York: A. A. Knopf, 2004.

In all of these broad phases, U.S. armored and tank-destroyer units played a role, although rarely the one envisioned by their designers. The experience of the forces that landed in General Omar Bradley's 12th Army Group is where the U.S. Army tanks and tank destroyers met a different combat reality than had been anticipated by prewar planners.

After landing in Normandy, the Allied offensive bogged down in the hedgerows of Normandy's *bocage* country. Historian Christopher Gabel's description of the fighting is illuminating:

> Seven weeks of bloody, grinding, attrition warfare followed. The terrain in the Normandy *bocage* consisted of small fields, roughly two hundred yards on a side, walled in by hedgerows—banks of earth topped by dense vegetation. The hedgerows provided the Germans with ready-made, compartmentalized defensive battlefield. German troops placed their automatic weapons, mortars, artillery, and antitank guns in concealed positions that afforded interlocking fields of fire across the small open spaces. Standard American fire-and-maneuver tactics fared poorly against such dug-in and camouflaged strong points. It was hard to locate the German positions; hence suppressive fires were generally ineffective. The hedgerows themselves restricted maneuver because any attempt to outflank a German position involved crossing a hedgerow and coming under fire from yet another enemy strong point. Engagements usually degenerated into costly infantry frontal assaults against concealed German machine guns and mortars. Infantry divisions and their attached tank battalions bore the brunt of the fighting on this terrible battlefield.[26]

Following the late-July 1944 breakout during Operation Cobra, the U.S. Army, spearheaded by its armored divisions, was able to shift into the exploitation and pursuit roles for which it was designed. This offensive stalled when the Allies overextended their logistics and when they ran into the Siegfried Line. The war in the west became a grueling battle of attrition against determined German defenders. Ironi-

---

[26] Gabel, "World War II Armor Operations in Europe," in Hofmann and Starry, eds., *From Camp Colt to Desert Storm*, p. 161.

cally, the Germans themselves caused the irretrievable weakening of their own defensive capacity when they spent their armies in the west during the abortive offensives in the Ardennes and Vosges in the winter of 1944–1945. In aftermath of these German last-gasp efforts, the Allies eventually returned to the offensive and, following the capture of a bridge at Remagen across the Rhine River in March 1945, swept through Germany in a final push that ended with the collapse of Nazi Germany in May 1945.

The U.S. Army's experience with armored forces began in World War I. During the Great War, the U.S. Army created a Tank Corps and fielded units that saw combat. The 304th Tank Brigade, under the command of Lieutenant Colonel George S. Patton, Jr., participated in the September 1918 offensive to reduce the St. Mihiel salient and in the final Allied Meuse-Argonne offensive that began on September 26, 1918, and ended with the Armistice in November. The U.S. components of the 304th consisted of the 344th and 345th Light Tank Battalions, which were equipped with French-supplied 7.4-ton Renault tanks. The other U.S. unit, the 301st Heavy Tank Battalion, was assigned to the British 2nd Tank Brigade. This battalion, equipped with British Mark V heavy tanks, supported the II American Corps in its assault on the Hindenburg line in late September 1918.[27]

The wartime experience of the U.S. Army Tank Corps was mixed. World War I tanks were slow, mechanically unreliable, and traversed the shell-pocked battlefields with difficulty; many broke down or became "ditched" in action. The 304th Tank Brigade began the Meuse-Argonne offensive with 142 Renault tanks. By November 1, 1918, it could field only 16 tanks to support the final assault.[28]

In the aftermath of World War I, U.S. Army reorganization legislation abolished the Tank Corps, largely because U.S. Army senior leadership, most notably General John J. Pershing, viewed the tank as an infantry support weapon. Leadership therefore believed that future

---

[27] Johnson, *Fast Tanks and Heavy Bombers*, pp. 36–37.

[28] Johnson, *Fast Tanks and Heavy Bombers*, p. 35. See also Dale E. Wilson, *Treat 'Em Rough! The Birth of American Armor, 1917–1920*, Novato, Calif.: Presidio Press, 1989, for a discussion of the World War I U.S. Army Tank Corps and its demise.

development of the tank should be left to infantry. Under the provisions of the National Defense Act of 1920, the newly constituted infantry branch was given responsibility for tanks, including the promulgation of doctrine and the establishment of materiel requirements. In 1931 the cavalry branch received War Department authority to develop tanks, which were known as "combat cars" to avoid the strictures of the National Defense Act.

Three critical factors affected the development of U.S. tanks: branch parochialism, weight constraints, and competing ideas about how to defeat enemy armor. Branch parochialism resulted in two categories of tanks: those designed as infantry support weapons, and those used as "iron horses" focused on traditional cavalry missions. Weight constraints affected what could be accomplished in tank design within the competing demands of speed, lethality, and protection. An increase in any of the three caused an increase in weight. A U.S. Army–imposed weight ceiling required trade-offs in one of the other three. Initially, the U.S. Army chief of engineers set the maximum weight for tanks at 15 tons, selecting this limit to match the carrying capacity of the divisional pontoon bridge.[29] Throughout the interwar period the U.S. Army Ordnance Department struggled unsuccessfully to provide the infantry and cavalry with tanks and combat cars that met their requirements within this weight limitation.[30] The U.S. Army increased tank weight limits later on, but still set a maximum (of 30 tons, with a width of 103 inches) to facilitate shipping and to ensure "that navy transporters and portable bridges did not need to be redesigned in the midst of the war."[31]

The question of how to use armored forces was also an issue in the U.S. Army. The chief of infantry believed that tanks existed to support attacking infantry. The 7th Cavalry Brigade (Mechanized), stationed at

[29] Johnson, *Fast Tanks and Heavy Bombers*, p. 74

[30] Johnson, *Fast Tanks and Heavy Bombers*, pp. 80, 200–201.

[31] House, *Combined Arms Warfare in the Twentieth Century*, p. 152. See also Baily, *Faint Praise*, p. 127. The 30-ton weight requirement in AR 850-15 was not relaxed until late 1944, when it was waived for the fielding of the T-26 tank. Because of its weight, the T-26 could not cross U.S. Army tactical bridges then in the field.

Fort Knox, Kentucky, willingly embraced armor, viewing combat cars as "iron horses." This brigade believed that tanks should be used, like traditional cavalry, to exploit and pursue infantry breakthrough attacks to complete the defeat of an enemy in depth. The chief of the Army Ground Forces, Lieutenant General Lesley J. McNair, was convinced that the appropriate response to the massed armored attacks employed in the German blitzkrieg offensives at the beginning of World War II was the tank destroyer.

Not until June 1940, after the success of the German blitzkrieg in Poland and France, did the U.S. Army merge its existing infantry and cavalry armored units into an Armored Force. The first chief of the Armored Force was Brigadier General Adna R. Chaffee, Jr., commander of the 7th Cavalry Brigade (Mechanized). At the critical period of the initial formation of the Armored Force and the armored division, General Chaffee shaped the armored divisions in the mold of the 7th Cavalry Brigade.[32]

During World War II, the U.S. Army fielded 16 armored divisions in Europe. Additionally, the U.S. Army put 65 independent tank battalions into the field to support its infantry divisions, compared to the 54 tank battalions within its armored divisions.[33] The U.S. Army had also fielded 61 tank-destroyer battalions in Italy and western Europe by war's end.[34] These forces were equipped with the weapons shown in Table 2.2.

Doctrine for U.S. Army tanks and tank destroyers evolved from the creation of the Armored Force in 1940 and the Tank Destroyer Center in 1941.[35] In both cases, steady institutional evolution as well as bottom-up innovation by units in the field throughout the war contributed to the development of doctrine. There were in essence two tank doctrines in the U.S. Army, each reflecting the interwar influence of

---

[32] Johnson, *Fast Tanks and Heavy Bombers*, p. 146.

[33] Gabel, "World War II Armor Operations in Europe," in Hofmann and Starry, eds., *From Camp Colt to Desert Storm*, p. 155.

[34] Gabel, "World War II Armor Operations in Europe," in Hofmann and Starry, eds., *From Camp Colt to Desert Storm*, p. 178.

[35] Baily, *Faint Praise*, p. 16.

infantry and cavalry perspectives on how best to employ armor. Thus, the U.S. Army fielded tank units to support the infantry divisions and serve as the principal units in armored divisions.

The nondivisional independent tank battalions were organized to support the mission that was aiding the advance of the infantry. The 1940 edition of FM 100-5, *Operations*, specified that these battalions would attack in two echelons. The first echelon sought to destroy enemy antitank guns; the second provided support to attacking infantry.[36] Cooperation between these tank battalions and supported infantry divisions continually improved because of their habitual association. It became standard practice to assign a tank battalion to infantry divisions and, generally, the companies of the battalion were spilt up to support the division's infantry regiments. Thus, "[f]or all practical purposes the tank company became an organic part of the infantry regiment."[37]

The formative doctrine of the early armored divisions derived from vintage cavalry doctrine and focused on "dash and speed rather than combined arms."[38] The March 1942 *Armored Force Field Manual* specified that the role was

> the conduct of highly mobile ground warfare, primarily offensive in character, by self-sustaining units of great power and mobility, composed of specially equipped troops of the required arms and services.[39]

The doctrine "was predicated on the assumption that tanks operated in masses, at their own pace, and that combined arms consisted of

---

36 Johnson, *Fast Tanks and Heavy Bombers*, pp. 145–146.

37 Gabel, "World War II Armor Operations in Europe," in Hofmann and Starry, eds., *From Camp Colt to Desert Storm*, pp. 162–163.

38 Gabel, "World War II Armor Operations in Europe," in Hofmann and Starry, eds., *From Camp Colt to Desert Storm*, p. 146.

39 Gabel, "World War II Armor Operations in Europe," in Hofmann and Starry, eds., From Camp Colt to Desert Storm, p. 147.

attaching supporting, subordinate elements to armored regiments."[40] Consequently, "[a]rtillery and infantry were subordinated to supporting roles—fixing the enemy and occupying captured positions."[41] The manual further "emphasized surprise, speed, shock action, and firepower directed against rear areas," and its "preferred tactics for armored formations were breakthrough, exploitation, encirclement, annihilation, and pursuit."[42] General McNair, in a January 23, 1943, memorandum to Army Chief of Staff General George C. Marshall, wrote that the "general concept" for the U.S. armored force was as "an instrument of exploitation, not greatly different in principle from horse cavalry of old."[43]

As the war progressed, the U.S. Army applied lessons learned from the field, particularly North Africa, and the doctrine for U.S. armored divisions took on more of a combined-arms tone. Thus, the January 1944 version of FM 17-100, *Armored Command Field Manual: The Armored Division*, "stressed the need for timely cooperation among the arms while placing more emphasis on the destruction of enemy forces in contact and less on cavalry-like rampages in hostile rear areas."[44] But combined arms in this manual referred to U.S. Army Ground Force units and did not envision the integration of air power to the levels realized in the German blitzkrieg. Furthermore, even if envisioned, air-ground cooperation would have been problematic, given the reality that the U.S. Army Air Forces were focused on strategic bombing and, as an institution, were not keen on the idea of subordinating air power

[40] Gabel, "World War II Armor Operations in Europe," in Hofmann and Starry, eds., *From Camp Colt to Desert Storm*, p. 143.

[41] Gabel, "World War II Armor Operations in Europe," in Hofmann and Starry, eds., *From Camp Colt to Desert Storm*, p. 149.

[42] Gabel, "World War II Armor Operations in Europe," in Hofmann and Starry, eds., *From Camp Colt to Desert Storm*, p. 147.

[43] Commander, Army Ground Forces, "Memorandum for CSA," January 23, 1943, cited in Kenneth A. Steadman, *The Evolution of the Tank in the U.S. Army*, Fort Leavenworth, Kan.: Combat Studies Institute, U.S. Army Command and General Staff College, 1982.

[44] Gabel, "World War II Armor Operations in Europe," in Hofmann and Starry, eds., *From Camp Colt to Desert Storm*, p. 147.

to ground forces. Instead, ad hoc procedures developed in the combat theaters to provide air support to ground units.[45]

Finally, one glaring deficiency in U.S. armored doctrine, both in the independent tank battalions and the armored divisions, was the assumption that tanks would not fight enemy tanks:

> "The main purpose of the tank cannon is to permit the tank to overcome enemy resistance and reach vital rear areas, where the tank machine guns may be used most advantageously."[46]

Therefore, although there was recognition that "[c]hance encounters between tanks would occur . . . the principal role of the armored division was to exploit and pursue, not to fight enemy armor."[47] If required, "antimechanized protection" would be provided by attaching tank destroyer units.[48]

Unfortunately for U.S. tankers, and in spite of U.S. Army doctrine, U.S. tanks did have to fight German tanks and did so at a great disadvantage. Most tank engagements were small actions. Historian Charles Baily notes that the 2nd Armored Division's biggest tank battle through the end of World War II "involved only twenty-five German tanks."[49] This action occurred in mid-November 1944 in the vicinity of Puffendorf, Germany. Over a two-day period, the U.S. 1st Battalion, 67th Armored Regiment, 2nd Armored Division, suffered 363 casualties and lost 57 tanks to well-sited German tanks. The battalion claimed only four German tanks destroyed—two by Shermans and two by M36 tank destroyers.[50]

U.S. Army doctrine envisioned two principal roles for tank destroyers. First, they supported offensive operations by protecting

[45] Johnson, *Fast Tanks and Heavy Bombers*, p. 226.

[46] Johnson, *Fast Tanks and Heavy Bombers*, p. 226.

[47] House, *Combined Arms Warfare*, p. 152.

[48] U.S. War Department, FM 17-100, *(Tentative) Employment of the Armored Division and Separate Armored Units*, pp. 1, 8, 13.

[49] Baily, *Faint Praise*, p. 92.

[50] Johnson, *Fast Tanks and Heavy Bombers*, p. 195.

friendly forces from enemy armored counterattacks. Second, they supported defensive operations by defending in depth against enemy armor attacks, with the majority of tank destroyers retained in a mobile reserve to respond to the main enemy attack.[51] This latter role was the main U.S. response to the German blitzkrieg.

In only one instance during World War II did a U.S. tank destroyer battalion ever execute its prescribed doctrine. During a March 1943 engagement near El Guettar in North Africa, the 601st Tank Destroyer Battalion (composed of M3 tank destroyers), with an attached company from the 899th Tank Destroyer Battalion (composed of M10 tank destroyers), turned back a German force of some 50 *Panzers*, but with heavy losses: 20 of 28 M3s and seven of ten M10s were destroyed.[52]

In northwest Europe, U.S. forces rarely encountered large German armored formations. Instead, the norm was the tough business of "[p]rying German infantry and guns from well-prepared positions."[53] In practice, much like the independent tank battalions, tank-destroyer battalions were semipermanently assigned to U.S. divisions and their companies were task-organized with infantry regiments.[54]

The types of maneuver anticipated by U.S. doctrine rarely occurred in northwest Europe. The most notable exception was the breakout from the Normandy beachhead and its exploitation during Operation Cobra in July 1944. Allied carpet bombing created a rupture in the German lines and U.S. ground forces broke out. Over the next several weeks the Allies swept through northern France and Belgium, and the "six American armored divisions in theater during the pursuit were

---

[51] Johnson, *Fast Tanks and Heavy Bombers*, p. 150.

[52] Gabel, "World War II Armor Operations in Europe," in Hofmann and Starry, eds., *From Camp Colt to Desert Storm*, p. 152.

[53] Baily, *Faint Praise*, p. 114.

[54] Gabel, "World War II Armor Operations in Europe," in Hofmann and Starry, eds., *From Camp Colt to Desert Storm*, p. 163.

in their glory."[55] However, the exploitation ground to a halt in early September:

> Logistical overextension, geography, weather, and a resurgent German defense all combined to stop the dash toward Germany. For the next several months the Allies were compelled to slog their way through forest, cities, and the fortifications of the Westwall.[56]

The tank destroyer battalions never maneuvered as had been envisioned; they were farmed out to the divisions to augment their combat power. The Germans executed only one massed armored attack, which was the very threat that tank destroyers had been created to counter. This massed armored attack occurred during the winter 1944 Ardennes offensive, and the tank destroyer battalions were largely dispersed throughout the 12th Army Group and incapable of executing the coordinated, mass maneuver called for in doctrine.

At the lowest tactical levels, the modern definition of maneuver—"systems move to gain positions of advantage against enemy forces"[57]—took on a particular importance because of the lethality and survivability advantage heavy German armor enjoyed over U.S. medium armor. Major General Maurice Rose, commander of the 3rd Armored Division, wrote about this problem in a March 1945 letter to General Eisenhower: "It is my personal conviction that the present M4A3 tank is inferior to the German Mark V." He told General Eisenhower that he had seen "projectiles fired by our 75mm and 76mm guns bouncing off the front plate of Mark V tanks at ranges of about 600 yards." U.S. tank crews had to close the range or angle for flank or rear shots, which was not always possible "due to the canalizing of the avenue

---

[55] Gabel, "World War II Armor Operations in Europe," in Hofmann and Starry, eds., *From Camp Colt to Desert Storm*, p. 167.

[56] Gabel, "World War II Armor Operations in Europe," in Hofmann and Starry, eds., From Camp Colt to Desert Storm, p. 169.

[57] U.S. Department of the Army, FM 3-0: *Operations*, p. 5-16.

of approach of both the German and our tank, which did not permit maneuver."[58]

The U.S. Army enjoyed two significant fire-support advantages over the German Army during the campaign in Europe: field artillery and air power. During the period between the two World Wars, the U.S. Army had developed fire-direction procedures that enabled multiple units to mass their fires on targets. Both air and ground forward observers who were linked to firing units through radios or wire communications provided responsive fires. Self-propelled and motorized field artillery provided a much higher degree of mobility than had been the case in previous wars.[59] Additionally, organizational arrangements, most notably the artillery group,

> permitted commanders to move artillery battalions from army to army, corps to corps, or division to division with ease and furnish additional artillery support where it was needed.[60]

The U.S. air-ground system was largely developed during combat operations. The U.S. Army Air Forces had played a key but largely independent role in strategic bombing and had made the invasion possible by isolating the Normandy beachhead.[61] Nevertheless, there was no system in place for supporting ground forces with air power when the invasion occurred.

The system of air-ground cooperation rapidly evolved after the invasion, largely through the initiative of Major General Elwood "Pete" Quesada, commander of the IX Tactical Air Command, which was supporting General Bradley's First Army. General Quesada collocated

---

[58] Major General Maurice Rose to General Dwight D. Eisenhower, March 21, 1945, cited in Johnson, *Fast Tanks and Heavy Bombers*, p. 199.

[59] Boyd L. Dastrup, *King of Battle: A Branch History of the U.S. Army's Field Artillery*, Fort Monroe, Va.: Office of the Command Historian, U.S. Army Training and Doctrine Command, 1993, p. 226.

[60] Dastrup, *King of Battle*, p. 220.

[61] Niklas Zetterling, *Normandy 1944: German Military Organization, Combat Power and Organizational Effectiveness*, Winnipeg, Manitoba: J. J. Fedorowicz Publishing, 2000, p. 112.

his operations center with that of Bradley's First Army,[62] and his efforts steadily paid dividends. During its pursuit following the Normandy breakout, the 4th Armored Division availed itself of this evolving system:

> Ninth Air Force's XIX Tactical Air Command provided a four-ship armed reconnaissance flight over each column. The fighter-bomber pilots, who were in direct radio communication with the commanders of the 4th Armored's combat commands, warned the ground elements of obstacles and enemy strong points, many of which they were able to neutralize before the armored columns arrived.[63]

Air power also helped make up for the disparity between German and U.S. armored vehicles. General Rose informed General Eisenhower in March 1945 that U.S. soldiers compensated for their "inferior equipment by the efficient use of artillery, air support, and maneuver."[64] Sergeant Harold E. Fulton perhaps said it best: "Our best tank weapon, and the boy that has saved us so many times, is the P-47 [fighter airplane]."[65] At the end of the war, the after-action review of the 12th Army Group noted emphatically the power of integrated ground and air power: "The air-armor team is a most powerful combination in the breakthrough and exploitation. . . . The use of this coordinated force, in combat, should be habitual."[66]

---

[62] Thomas Alexander Hughes, *Over Lord: General Pete Quesada and the Triumph of Tactical Air Power in World War II*, New York: Free Press, 1995, pp. 156–158.

[63] Gabel, "World War II Armor Operations in Europe," in Hofmann and Starry, eds., *From Camp Colt to Desert Storm*, p. 169.

[64] General Rose to General Eisenhower, cited in Johnson, *Fast Tanks and Heavy Bombers*, p. 199.

[65] Brigadier General I. D. White to General Eisenhower, March 20, 1945, cited in Johnson, *Fast Tanks and Heavy Bombers*, p. 199.

[66] U.S. 12th Army Group, *12th Army Group Report of Operations (Final After Action Report)*, Vol. 11, *Antiaircraft Artillery, Armored, Artillery, Chemical Warfare and Signal Sections*, 1945, p. 61. See also The General Board, United States Forces, European Theater, *The Tactical Air Force in the European Theater of Operations*, Study Number 54, circa 1946, Foreword, p. 1. This study notes the importance of air power to ground operations:

Key Insights

World War II was the first war in which large-scale armored warfare played a major role. The war also yielded important lessons about armored vehicles and doctrine that would influence the U.S. Army in the coming decades:

- U.S. armored doctrine, and the tanks and tank destroyers fielded to support it, reflected internal U.S. Army institutional preferences and agendas, particularly those of the infantry and cavalry branches. The tank destroyer and its doctrine were imposed as a response to antiarmor warfare by General McNair. Neither was an informed response to German capabilities. In northwest Europe in 1944 and 1945, U.S. armored doctrine and equipment proved to be deeply flawed. In the aftermath of the war, the U.S. Army recognized these flaws, abandoned tank destroyers, and emphasized the need for lethality and survivability as key features in future tanks. The report of a postwar general board was quite clear in this regard: "[T]he European campaign demonstrated that tanks fight tanks." For future tank guns, the board recommended a "*minimum* standard" for "exploitation tanks" in an armored division: a "gun capable of penetrating the sides and rear of any enemy armored vehicle and the front of any but the heaviest assault tank." Future U.S. tank survivability also received attention from the board: "Frontal armor and armor over ammunition stowage must be capable of withstanding all foreign and anti-tank weapons at normal combat ranges."[67] Every U.S. tank since World War II has reflected these design criteria, with

---

> The entrance of ground components was timed on air capabilities. Air success was so important that if delays occurred due to weather, or other causes, ground action on an Army or larger scale was postponed.

However, War Department doctrine was, in the view of this report, inadequate in the realm of ground-air doctrine, and procedures were largely ad hoc: "No manual existed except Field Manual 31-35 ... which had been obsolete for years. The splendid cooperation between the Tactical Air Commands and the Armies was developed during operations."

[67] The General Board, United States Forces, European Theater, *Tank Gunnery*, Study Number 53, circa 1946, p. 29, and The General Board, United States Forces, European The-

the M1 Abrams tank displaying the ultimate evolution of these principles.

- U.S. tanks and tank destroyers, although at a disadvantage in direct tank-on-tank combat against more-lethal and more-survivable German armored vehicles, served two important functions. First, they provided platforms that could maneuver with armored protection and speed not possible in the U.S. Army's infantry divisions. Second, they served as valuable fire-support platforms for infantry regiments.
- The U.S. Army focus on deployability and battlefield mobility constrained tank development by imposing weight limitations on rapidly evolving armored system technologies. These weight limitations were most significant in the areas of lethality and survivability.
- Tanks were most effective when employed in a combined-arms context that incorporated infantry, field artillery, and air power. The tactics, techniques, and procedures (TTP) for U.S. combined-arms warfare, however, evolved mainly in the field because existing doctrine, organizational arrangements, and training were largely inadequate.
- U.S. armored vehicles were highly vulnerable to German anti-armor weapons that ranged from individual short-range weapons (e.g., *panzerfaust*) to tanks and tank destroyers. Furthermore, German defenses were generally integrated and resilient and posed significant challenges to the execution of U.S. doctrinal concepts. These vulnerabilities were heightened in complex terrain (e.g., *bocage* and urban settings).

---

ater, *Organization, Equipment, and Tactical Employment of Separate Tank Battalions,* Study Number 50, circa 1946, p. 12.

U.S. Army Armored Cavalry and Mechanized Infantry in Vietnam (1965–1972)

U.S. involvement in Vietnam dates back to World War II,[68] when Office of Strategic Services (OSS) operatives provided modest support to Ho Chi Minh's Viet Minh in their war against Japan.[69] At the end of the war, the United States provided military aid to France as it attempted to reassert control over its colonial possessions in Indochina—control that was contested by the Viet Minh. The first U.S. advisors arrived in Saigon in August 1950. The French effort continued until 1954, when the loss at Dien Bien Phu, after a 55-day siege, convinced the French of the inevitability of withdrawal from Vietnam.[70]

The 1954 Geneva Conference on Indochina ended the fighting and established a military demarcation line that temporarily divided Vietnam into northern and southern parts at the 17th parallel. Elections were scheduled for 1956 to choose a government to lead a reunified Vietnam.

U.S. President Dwight D. Eisenhower began the formal U.S. relationship with South Vietnam when he included South Vietnam in the Southeast Asia Treaty Organization in September 1954. In October President Eisenhower pledged U.S. support to South Vietnamese Prime Minister Ngo Dinh Diem. In 1955, Prime Minister Diem became president of the independent Republic of Vietnam. The 342 U.S. advisors in the U.S. Military Assistance Advisory Group, Vietnam, began training, organizing, and equipping the South Vietnamese Army to fend off an invasion from the North.

---

[68] The literature on Vietnam War is extensive. Excellent secondary sources are: Guenter Lewy, *America in Vietnam*, New York: Oxford University Press, 1978; George C. Herring, *America's Longest War: The United States and Vietnam, 1950–1975,* 3rd ed., New York: McGraw-Hill, 1996; and Bruce Palmer, *The 25-Year War: America's Military Role in Vietnam*, New York: Touchstone, 1984.

[69] Central Intelligence Agency, "OSS in Asia," Center for the Study of Intelligence Publications, November 19, 2007.

[70] Lewy, *America in Vietnam,* p. 7. President Eisenhower refused to initiate active U.S. intervention to relieve the siege of Dien Bien Phu.

In 1956 President Diem, supported by the United States, refused to hold the elections required by the Geneva Conference. He also began a campaign to eliminate his political opponents and the Viet Minh, who became known as the Viet Cong (VC). North Vietnam actively supported the growing insurgency and created the National Liberation Front in 1960 to further the cause of unifying Vietnam. VC ranks swelled from some 5,000 in 1960 to approximately 100,000 by 1964.

The United States responded to the growing insurgency with direct military involvement in 1961, when it deployed combat advisors. By July 1963, there were 15,400 U.S. troops in South Vietnam.

The United States reached a critical juncture in its involvement after an alleged August 1964 attack on two U.S. destroyers by North Vietnamese patrol boats in the Gulf of Tonkin. Congress gave U.S. President Lyndon B. Johnson broad authority in the Gulf of Tonkin Resolution, "authorizing all actions necessary to protect American forces and to provide for the defense of the nation's allies in Southeast Asia."[71] President Johnson authorized air strikes against North Vietnam, and in March 1965 U.S. Marines deployed to South Vietnam. Several weeks later, the first U.S. Army ground combat unit, the 173rd Airborne Brigade, arrived in South Vietnam, raising overall U.S. military strength in the country to over 50,000. In July 1965 President Johnson announced further deployments that would increase the U.S. presence to 175,000. The U.S. Army's 1st Cavalry Division (Airmobile), the 1st Brigade of the 101st Airborne Division, the 1st Infantry Division, and numerous support units deployed to Vietnam as part of this buildup.[72] The scheduled deployment of the 1st Infantry Division, which contained mechanized infantry and tank battalions, forced the U.S. Army to address the question of how it would use its armored forces in Vietnam.

---

[71] Maurice Matloff, ed., *American Military History*, rev. ed., Washington, D.C.: Office of the Chief of Military History, U.S. Army, 1989, p. 637.

[72] At the height of the Vietnam War, the United States had over 500,000 troops deployed in Vietnam.

The Armored Forces

When the United States began deploying significant ground forces to Vietnam in 1965, prevailing wisdom in the U.S. Army was that the conflict was "an infantry and Special Forces fight."[73] This view persisted at the highest levels. General Harold K. Johnson, the Army Chief of Staff in 1965, and General William C. Westmoreland, the commander of the U.S. Military Assistance Command, Vietnam (MACV), both believed that armor had little utility in Vietnam.[74]

Generals Johnson and Westmoreland had a mix of reasons for their opinions about armor in Vietnam. First, as the U.S. Army began its massive expansion to pursue a search-and-destroy strategy in Vietnam, it faced troop ceilings. Obviously, given these ceilings, U.S. Army leadership wanted to deploy the most effective force possible within the manpower constraints. Second, if armored units did deploy, they would require proportionally more support troops and a more sophisticated logistical base to support them than would infantry units. The generals' belief that the utility of armor in the Vietnam conflict would be limited also derived from less objective sources. These surfaced when the 1st Infantry Division was preparing to deploy to Vietnam in the summer of 1965.[75]

The 1st Infantry Division was reorganized to prepare for Vietnam. The Army eliminated the division's two tank battalions and dismounted its mechanized infantry battalions. General Johnson's predispositions about the utility of armor in Vietnam became clear when he responded to an Army Staff recommendation that one of the division's tank battalions go with it to Vietnam. General Johnson disapproved the request, noting four principal reasons. First, he alluded to the Korean War experience where "the oriental" had effectively used mines against armor and where U.S. tanks "had limited usefulness." He expected to encounter the same mine problem in Vietnam. Thus, General Johnson noted that

---

[73] Donn A. Starry, *Mounted Combat in Vietnam*, Washington, D.C.: Department of the Army, 1989, p. 51.

[74] Hofmann and Starry, eds., *From Camp Colt to Desert Storm*, pp. 325–326.

[75] Hofmann and Starry, eds., *From Camp Colt to Desert Storm*, p. 55.

> [o]n balance, in Vietnam the vulnerability to mines and the absence of major combat formations in prepared positions where the location is accessible lead me to the position that an infantry battalion will be more useful to you than a tank battalion, at this stage.

Second, General Johnson wrote that he was not aware that the South Vietnamese found much use for the light tanks in their army. Third, General Johnson believed that tanks would slow down the "rapid movement of troops." Finally, he evoked the French experience: "The presence of tank formations tends to create a psychological atmosphere of conventional combat, as well as recalls the image of French tactics in the same area in 1953 and 1954." General Johnson's only concession was to allow the 1st Infantry Division's armored cavalry squadron (1st Squadron, 4th Cavalry) to retain its M48A3 medium-armored tanks and M113 armored personnel carriers (APCs) in order "to test the effectiveness of armor," although he added that he would revisit his decision if circumstances changed.[76]

General Westmoreland fully supported General Johnson's decision not to deploy the majority of the 1st Infantry Division's armored vehicles. He assured General Johnson that, "except for a few coastal areas, most notably in the I Corps area, Vietnam is no place for either tank or mechanized infantry units."[77] Ironically, much of the U.S. Army's armor community agreed with Generals Johnson and Westmoreland. Indeed, "many senior armor officers who had spent years in Europe dismissed the Vietnam conflict as a short, uninteresting interlude best fought with dismounted infantry."[78]

When the 1st Infantry Division arrived in Vietnam, the three armored cavalry troops of the 1st Squadron, 4th Cavalry, were parceled out to the division's three brigades. Due to a "'no tanks in the jungle'

---

[76] Quotes are from a July 3, 1965, message from General Harold K. Johnson to General William C. Westmoreland, cited in Hofmann and Starry, eds., *From Camp Colt to Desert Storm*, pp. 55–56.

[77] Hofmann and Starry, eds., *From Camp Colt to Desert Storm*, p. 56.

[78] Hofmann and Starry, eds., *From Camp Colt to Desert Storm*, p. 51.

attitude," the squadron's M48A3 tanks were completely pulled out for six months, not returning until the division and squadron commanders convinced General Westmoreland that tanks were indeed useful. Meanwhile, the division's two tank battalions, left at Fort Riley, were still withheld.[79]

Nevertheless, by the end of 1965, as the United States poured more forces into Vietnam in the face of a deteriorating military situation, U.S. armored forces began arriving in South Vietnam. The 25th Infantry Division brought its tank battalion, mechanized infantry battalion, and armored cavalry squadron, largely because of the insistence of its commander, Major General Frederick C. Weyand. The 11th

---

[79] Hofmann and Starry, eds., *From Camp Colt to Desert Storm*, p. 57. On the origins of the aversion of many U.S. Army officers to armor in Vietnam, see also pp. 4–5, where the authors note the following:

> In the United States, because of restrictive military security regulations and a general lack of interest in the French operation in Indochina, there was no body of military knowledge of Vietnam. What was known had been drawn not from after action reports but from books written by civilians. Foremost among these was Bernard B. Fall's *Street Without Joy*, which greatly influenced the American military attitude toward armored operations in Vietnam. One series of battles in particular stood out from all the rest, epitomized the French experience in American eyes. Entitled "End of a Task Force," Chapter 9 of Fall's widely read book traced a six-month period in the final struggles of a French mobile striking force, Groupement Mobile 100. The vivid and terrifying story of this group's final days seemed to many to describe the fate in store for any armored unit that tried to fight insurgents in the jungle.

The authors also note that General Westmoreland "kept a copy of Fall's book on the table near his bed. He later said that the defeat of Groupement Mobile 100 was 'always on my mind,' particularly so during the early U.S. deployments." See also, p. 327, where General Michael S. Davison recalls difficulties in convincing MACV to deploy the 11th Armored Cavalry to Vietnam:

> [W]e had a hell of a time selling that one, not only had difficulty selling it, but difficulty in selling the use of tanks at all in any form in Vietnam, because General Westmoreland and Bill DePuy, who was his J3 in this period, couldn't conceive of tanks or armored cavalry being able to do anything in Vietnam. . . . [T]his is a factor of, really, their own lack of experience with armor.

For a more recent discussion of Groupement Mobile 100, see Kirk A. Luedeke, "Death on the Highway: The Destruction of Groupement Mobile 100," *Armor*, Vol. 110, No. 1, January–February 2001, pp. 22–29.

Armored Cavalry Regiment, with its tanks and APCs, was also alerted for deployment.[80]

One of the major factors that had influenced MACV to oppose the deployment of armored units was the belief that the environment in Vietnam was not conducive to armored operations. This view was largely dispelled in the 1967 "Evaluation of U.S. Mechanized and Armored Combat Operations in Vietnam (MACOV)" report. This report assessed terrain and weather conditions in Vietnam and found that tanks and APCs could negotiate many areas of the country during both the dry and monsoon seasons.[81] The report also noted that,

> [c]ontrary to established doctrine, armored units in Vietnam were being used to maintain pressure against the enemy in conjunction with envelopment by airmobile infantry. Moreover, tanks and APCs frequently preceded rather than followed dismounted infantry through the jungle, where they broke trail, destroyed antipersonnel mines, and disrupted enemy defenses.[82]

As the U.S. Army's ground combat strength in Vietnam increased to execute MACV's search-and-destroy strategy, so too did the number of armored units. Eventually, one

> armored cavalry regiment, three tank battalions and a separate tank company, six armored cavalry squadrons, ten mechanized infantry battalions, twenty-two armored artillery battalions, and four armored cavalry troops [served in Vietnam].[83]

The principal armored vehicles used by the U.S. Army in Vietnam were the M48A3 tank, the M113 APC, and the M551 airborne assault/armored reconnaissance vehicle. Table 2.4 shows the characteristics of these vehicles.

---

[80] Starry, *Mounted Combat in Vietnam*, p. 58.

[81] Hofmann and Starry, eds., *From Camp Colt to Desert Storm*, pp. 327–328.

[82] Starry, *Mounted Combat in Vietnam*, p. 85.

[83] Hofmann and Starry, eds., *From Camp Colt to Desert Storm*, p. 325.

Table 2.4
U.S. Armored Vehicles in Vietnam

System	Weight (tons)	Armament	Max Armor (mm)
M113 armored cavalry assault vehicle (ACAV)	12.5	.50-caliber machine gun; two 7.62-mm machine guns	38
M551 Sheridan airborne assault/armored reconnaissance vehicle	17.4	152-mm gun and missile launcher; .50-caliber machine gun; 7.62-mm machine gun	N/A[a]
M48A3 tank	51.9	90-mm gun; .50-caliber machine gun; 7.62-mm machine gun	120

SOURCE: Christopher F. Foss, *Jane's World Armoured Fighting Vehicles*, New York: St. Martin's Press, 1976, pp. 89–91, 100–105, 294–296.

[a] Aluminum body; steel turret.

The M48A3 was a second-line U.S. tank. The more modern M60 tanks, armed with 105-mm main guns, were slated for potential European battlefields and never deployed to Vietnam. The M48A3 mounted a 90-mm main gun, a turret-mounted .50-caliber machine gun, and a 7.62-mm coaxially mounted machine gun. The 90-mm gun could fire high-explosive, high-explosive antitank, white phosphorous, and canister and "beehive" antipersonnel rounds.[84] The M48A3 could also be equipped with the infrared- and white-light–capable xenon searchlight that was useful in night operations. Given its size and power, M48A3s were "[o]ften used as mobile battering rams in 'jungle' busting operations."[85]

The M113 was the workhorse of the armored forces in Vietnam. With its amphibious capability and high degree of mobility, the M113 could be deployed almost anywhere in Vietnam. It was frequently modified into an armored cavalry assault vehicle (ACAV), following the example of South Vietnamese modifications to their M113s, by the addition of an armored shield to the .50-caliber machine gun and the mounting of two 7.62-mm M60 machine guns. Furthermore, M113s

[84] Beehive rounds contained hundreds of "flechettes," or small darts, that were very effective against enemy personnel.

[85] Hofmann and Starry, eds., *From Camp Colt to Desert Storm*, p. 329.

were used as fighting vehicles, contrary to their doctrinal role as an infantry carrier. According to doctrine, infantrymen were to dismount from the M113 to maneuver and fight. In Vietnam, soldiers often remained in the vehicle and fought from it.[86]

The M551 Sheridan airborne assault/armored reconnaissance vehicle first deployed to Vietnam in January 1969 when it replaced M48A3 tanks and some ACAVs in cavalry units. The M551 had a 152-mm main gun that could fire antitank guided missiles or conventional rounds with combustible cartridges, including high-explosive antitank, canister, and beehive rounds. Additionally, the M551 had a .50-caliber machine gun on the turret and a coaxially mounted 7.62-mm machine gun, and could be fitted with a xenon searchlight.[87]

Employment

U.S. armored formations, although organized and equipped for armored warfare in a conventional setting, adapted themselves to the operational environment, often in nondoctrinal ways. Armored cavalry units, in addition to executing their doctrinal missions of "reconnaissance, security, and economy of force" also performed "convoy escort, search and destroy (mounted and dismounted), cordon and search, search and clear, route clearing, and base defense reaction force" missions. As for tank units, General Harold K. Johnson, Army Chief of Staff, noted in 1967 that

> tank units are often tasked to link up with airmobile infantry. Tank units with attached infantry are also performing search and destroy, convoy escort, and security missions similar to those assigned to armored cavalry units. . . . Tank and armored cavalry units also provided protection for land clearing teams and supported the pacification program. On some occasions tanks were even employed in an indirect fire role, supplementing available artillery.

[86] Hofmann and Starry, eds., *From Camp Colt to Desert Storm*, pp. 330–332.

[87] Hofmann and Starry, eds., *From Camp Colt to Desert Storm*, pp. 333–334. Approximately 200 M551s were fielded in Vietnam by late 1970. The missile-firing capability of the M551 was never used in Vietnam.

Additionally, armored units "with the advent of airmobile infantry . . . were often used to fix the enemy while airmobile infantry deployed as the maneuver element." Later in the war,

> as pressure increased to hold down American casualties during disengagement, more and more combat elements of whatever type sought to fix the enemy when contact was made while firepower of every description was used for exploitation.[88]

Finally, the protected mobility of U.S. armored vehicles made them invaluable resources during the U.S. response to the January 1968 Tet Offensive and two smaller communist efforts in May and August 1968: "When the enemy forced free world forces to move rapidly from one battle area to another, it was the armored forces that covered the ground quickly and in many cases averted disaster."[89] The contribution of armored forces was particularly important in the major cities of South Vietnam. In Saigon, "cavalry and mechanized infantry decided the fate of the city."[90] In Hue, the fires provided by U.S. and South Vietnamese armored forces provided a crucial edge to infantry in the brutal 26-day battle to clear the city of communists.[91]

U.S. Army doctrine for the employment of U.S. armored forces focused on conventional combat in Europe.[92] As previously noted, U.S. armored forces largely abandoned these conventional concepts and adapted to the environment in which they found themselves. The U.S. Army Armor School and the Combat Developments Command Armor Agency at Fort Knox, Kentucky, rejected any doctrinal changes based on U.S. experience in Vietnam, arguing that "new concepts were not applicable to armor combat in other parts of the world."[93] Regard-

---

[88] Hofmann and Starry, eds., *From Camp Colt to Desert Storm*, p. 336.

[89] Starry, *Mounted Combat in Vietnam*, p. 115. This volume describes the various actions U.S. armored forces were involved in during the three communist offensives of 1968.

[90] Starry, *Mounted Combat in Vietnam*, p. 118.

[91] Starry, *Mounted Combat in Vietnam*, p. 116.

[92] Starry, *Mounted Combat in Vietnam*, p. 7.

[93] Starry, *Mounted Combat in Vietnam*, p. 86.

ing the use of M113s as fighting vehicles, the Continental Army Command (CONARC) believed that "adopting as doctrine the employment of mounted infantry in a cavalry role was neither feasible nor desirable."[94]

Key Insights

Although the Vietnam War did not conform to doctrinal norms for the use of armor envisaged by the United States, the case provides several insights about the use of medium-armored forces in counterinsurgency, complex terrain, and low- to mid-intensity conflict. The armored vehicles the that U.S. Army employed in Vietnam were medium-weight platforms—like the M113 and the Sheridan—or the second-tier M48 tank. The following key insights emerge:

- U.S. medium-armored vehicles were able to operate in Vietnam's complex terrain, including jungle and semimountainous territory. Tanks were somewhat less versatile.
- Armored forces effectively complemented light and air-mobile forces, both as maneuver elements and firepower platforms.
- Per the MACOV study, medium-weight vehicles (i.e., the M113) had greater trafficability than tanks in complex terrain (i.e., jungle and semimountainous territory). In cities, medium-weight vehicles provided protected mobility, rapid reaction, and fire support.
- Each type of armored vehicle employed by the United States in Vietnam was vulnerable to mines, improvised explosive devices (IEDs), rocket-propelled grenades (RPGs), and recoilless rifles. Enemy forces learned the vulnerabilities of armored vehicles and exploited them.
- Casualty rates among mechanized infantry units were lower than those of other light infantry units because of the armored mobility afforded by M113s.

---

94 Starry, *Mounted Combat in Vietnam*, p. 86.

Task Force Shepherd, 1st Marine Division, in Operations Desert Shield and Desert Storm (Southwest Asia, 1990–1991)

In the early hours of August 2, 1990, Iraq invaded Kuwait with three Republican Guard Divisions and special operations forces, encountering little resistance. By noon on the following day, Iraq occupied Kuwait City and key locations throughout Kuwait, and had begun positioning forces along the Kuwait–Saudi Arabia border. By August 6, Iraq was believed to have some 200,000 men and 2,000 tanks deployed in or around Kuwait.[95]

The international community's response was swift. On August 2, the United Nations (UN) Security Council passed Resolution 660, condemning the invasion and calling for the unconditional withdrawal of Iraqi troops from Kuwait. Resolution 661 followed on August 6, imposing sanctions and an embargo on Iraq.

The United States was also quick to act. On August 2, President George H. W. Bush issued Executive Orders 12722 and 12723, declaring a national emergency, imposing trade sanctions on Iraq, and freezing Iraqi and Kuwaiti financial assets. The Joint Staff and U.S. Central Command (CENTCOM) began reviewing and revising Operation Plan (OPLAN) 102-90, the war plan for the region, and devising plans for the defense of Saudi Arabia. On August 3, U.S. naval forces began deploying to Southwest Asia, and on August 4, General H. Norman Schwarzkopf, Jr., Commander, CENTCOM, and Lieutenant General Charles A. Horner (the CENTCOM air component commander) briefed a concept for the defense of Saudi Arabia to President Bush at Camp David.

On August 5, President Bush vowed that the Iraqi invasion of Kuwait would "not stand," and demanded a complete Iraqi withdrawal from Kuwait. This demand was central to the framework of U.S. objectives in the region after the Iraqi invasion of Kuwait. The August 6 articulation of the framework included the following objectives:

---

[95] Anthony H. Cordesman and Abraham R. Wagner, *The Lessons of Modern War:* Vol. IV, *The Gulf War*, Boulder, Colo.: Westview Press, 1996, pp. 47–49.

- immediate, unconditional, and complete withdrawal of all Iraqi forces from Kuwait
- restoration of Kuwait's legitimate government
- ensured stability and security of Saudi Arabia and the Persian Gulf
- ensured safety and protection of the lives of U.S. citizens abroad.[96]

President Bush also dispatched Secretary of Defense Richard Bruce "Dick" Cheney, General Schwarzkopf, and General Horner to Saudi Arabia to confer with Fahd bin Abdul Aziz Al Saud, King of Saudi Arabia. Secretary Cheney briefed King Fahd on August 6, assuring him that the United States was committed to defending Saudi Arabia. Secretary Cheney also presented plans to reinforce Saudi forces to this end. The Saudis accepted the plan, and on August 8, President Bush announced the deployment of U.S. forces to defend Saudi Arabia in an operation known as Desert Shield. [97]

One of the first units alerted for movement to Saudi Arabia was the 7th Marine Expeditionary Brigade (MEB), which was stationed at the Marine Corps Air Ground Combat Center, Twenty-Nine Palms, California. The 7th MEB, commanded by Major General John I. Hopkins, received deployment orders on August 10, and was destined for the port of Al Jubayl, Saudi Arabia. General Hopkins requested and received operational control of the 1st Marine Division's contingency forces stationed at Camp Pendleton, California. The mission assigned to the 7th MEB "was to prepare to protect critical oil and port facilities and delay any advancing Iraqi force as far north as possible."[98] At Al Jubayl, the personnel of the 7th MEB linked up with equipment from the Maritime Pre-Positioning Squadron (MPS) 2. By August 26,

---

[96] Cordesman and Wagner, *The Gulf War*, p. 53.

[97] "The Gulf War: A Chronology," *Air Force*, Vol. 84, No. 1, January 2001; Cordesman and Wagner, *The Gulf War*, pp. 50–53.

[98] Charles H. Cureton, *United States Marine Corps in the Persian Gulf, 1990–1991: With the 1st Marine Division in Desert Shield and Desert Storm*, Washington, D.C.: Headquarters, U.S. Marine Corps, 1993, p. 1.

the airlift of 7th MEB personnel was complete and General Hopkins deployed the unit, now designated the 7th Regimental Combat Team (RCT), in defensive positions north of Al Jubayl.[99]

The 1st MEB, MPS 3, and the remaining elements of the 1st Marine Division soon joined the 7th RCT in Saudi Arabia. As these units arrived, they were combined into the 1st Marine Division, commanded by Brigadier General James M. Myatt.[100] General Myatt focused initially on the original defensive mission assigned to the 7th MEB, but also prepared for the employment of the division as soon as possible to

> "attrit and delay" an advancing enemy. Other tasks included conducting close air support and interdiction operations, and planning counteroffensive operations to restore the integrity of the Saudi Arabian border [if it were violated].[101]

Furthermore, the division began the evolving process of incorporating arriving units into the division, refining itself into a combat organization whose

> ultimate aim was the eventual creation of task forces that had the mobility, fire power and engineer capability to penetrate Iraqi defensive lines and then defeat enemy mechanized and armored forces.[102]

Task Force Shepherd, named after former Marine Corps Commandant Lemuel C. Shepherd, Jr., and commanded during Operations Desert Storm and Desert Shield by Lieutenant Colonel Clifford O.

---

[99] Cureton, *United States Marine Corps in the Persian Gulf,* pp. 1–4. See also Charles D. Melon, Evelyn A. Englander, and David A. Dawson, *U.S. Marines in the Persian Gulf, 1990–1991: Anthology and Annotated Bibliography,* Washington, D.C.: History and Museums Division, Headquarters, U.S. Marine Corps, [1992] 1995, p. 130. General Hopkins told General Myatt that he was ready to defend on August 25.

[100] Cureton, *United States Marine Corps in the Persian Gulf,* pp. 8–9. General Myatt was frocked to Major General before Desert Storm.

[101] Cureton, *United States Marine Corps in the Persian Gulf,* p. 9.

[102] Cureton, *United States Marine Corps in the Persian Gulf,* p. 9.

Myers (commander, 1st Light Armored Infantry Battalion), was one of the task forces created within the 1st Marine Division. Although the composition of Task Force Shepherd changed several times during the Gulf War, at its core, it was a grouping of LAV-equipped, light-armored infantry units within the 1st Marine Division. [103]

The Armored Forces

Composed of various groupings of companies from the 1st Light Armored Infantry Battalion and the 3rd Light Armored Infantry Battalion, Task Force Shepherd was equipped with medium-armored vehicles when compared to the lightly armored amphibious assault vehicles (AAVs) and trucks used to transport U.S. Marine infantry and the M1 tanks within the 1st Marine Division. It was also medium-armored in the context of the Iraqi Army's equipment, which included tanks.[104] A description of the different LAV variants is provided in Table 2.5. The types of armored vehicles available to the Iraqi Army are displayed in Table 2.6.

Employment

Operation Desert Shield. During Operation Desert Storm, Task Force Shepherd conducted screening and minor reconnaissance missions, operating in a manner reminiscent of U.S. Army armored cavalry squadrons.[105] By the end of August 1990, Task Force Shepherd deployed in front of the 1st Marine Division and conducted long-range screening operations. Behind this screen, the 1st Marine Division received and incorporated deploying units. By mid-September, the division had built up sufficient combat strength for General Myatt to expand the division's mission. He deployed additional forces to the north "to pro-

---

[103] Cureton, *United States Marine Corps in the Persian Gulf*, pp. 9–10.

[104] See Cureton, *United States Marine Corps in the Persian Gulf*, p. 137. The 1st Marine Division had the following LAV systems in the following quantities: light-armored vehicle–reconnaissance/troop carrier (LAV-25), 73 units; light-armored vehicle–antitank (LAV-AT), 22 units; light-armored vehicle–logistic (LAV-L), 15 units; light-armored vehicle–mortar (LAV-M), ten units; light-armored vehicle–recovery (LAV-R), eight units; and light-armored vehicle–command and control (LAV-C2), 11 units.

[105] Hofmann and Starry, eds., *From Camp Colt to Desert Storm*, p. 487.

Table 2.5
Task Force Shepherd Medium Armor in Operations Desert Shield and Desert Storm

Type	Weight (tons)	Armament	Max Armor (mm)
LAV-25	14.100	25-mm chain gun; 7.62-mm machine gun	10
LAV-AT	13.825	antitank guided missile (ATGM); 7.62-mm machine gun	10
LAV-M	13.350	81-mm mortar; 7.62-mm machine gun	10
LAV-C2	13.530	7.62-mm machine gun	10
LAV-L	14.100	7.62-mm machine gun	10
LAV-R	14.160	7.62-mm machine gun	10

SOURCE: Christopher F. Foss, *Jane's Tanks and Combat Vehicles Recognition Guide*, New York: Harper Collins, 2000; U.S. Marine Corps, Headquarters, homepage, n.d.

tect Jubayl by disrupting, delaying, and destroying an attacking Iraqi force."[106] Task Force Shepherd's role was to screen the division along a 60-mile front. It also worked with the reinforced 3d Battalion, 9th Marines, to provide the division with a mobile rapid reaction force.[107]

On November 29, a critical threshold was crossed when the UN Security Council passed Resolution 678, authorizing the use of "all necessary means" to force Iraq out of Kuwait. The resolution set a deadline of January 15, 1991, for Saddam Hussein to comply with earlier UN resolutions.[108] The 1st Marine Division began preparing for offensive operations, with a "be-prepared date" of January 15, 1991. Task Force Shepherd continued its covering operation.[109]

The Battle for Observation Post 4. The January 15 deadline passed without an Iraqi withdrawal from Kuwait. Accordingly, coalition air

---

[106] Cureton, *United States Marine Corps in the Persian Gulf*, p. 12.

[107] Cureton, *United States Marine Corps in the Persian Gulf*, pp. 12–13.

[108] Cordesman and Wagner, *The Gulf War*, p. 52.

[109] Cureton, *United States Marine Corps in the Persian Gulf*, p. 13.

Table 2.6
Iraqi Armor in Operations Desert Shield and Desert Storm

Type	Country of Origin	Weight (Tons)	Armament	Max Armor (mm)
T-54/55 tank	Soviet Union	39.00	100-mm gun; 12.7-mm machine gun; 7.62-mm machine gun	203
T-62 tank	Soviet Union	44.00	115-mm gun; 12.7-mm machine gun; 7.62-mm machine gun	242
T-72 tank	Soviet Union	48.90	125-mm gun; 12.7-mm machine gun; 7.62-mm machine gun	N/A[a]
PT-76 reconnaissance vehicle	Soviet Union	16.06	7.62-mm gun; 7.62-mm machine gun	14
BMD-1 airborne combat vehicle	Soviet Union	8.20	73-mm gun; three 7.62-mm machine guns	23
BMP-1 IFV	Soviet Union	14.80	73-mm gun; 7.62-mm machine gun; ATGM	33
BMP-2 IFV	Soviet Union	15.70	30-mm cannon; 7.62-mm machine gun; ATGM	N/A[b]
BRDM-2 amphibious reconnaissance vehicle	Soviet Union	7.20	14.5-mm machine gun, 23-mm cannon, or ATGM; 7.62-mm machine gun	7[c]
BTR-50P/BTR-50PK APC	Soviet Union	15.60	7.62-mm machine gun	10
BTR-60PB APC	Soviet Union	11.40	14.5-mm machine gun; 7.62-mm machine gun	9
MT-LB APC	Soviet Union	13.09	7.62-mm machine gun	10
YW-531–series APC	China	13.90	12.7-mm machine gun	10
EE-9 Cascavel armored car	Brazil	14.80	90-mm gun; 7.62-mm machine gun or 12.7-mm machine gun	N/A[d]
EE-11 Urutu APC	Brazil	15.40	14.5-mm machine gun, 12.7-mm machine gun, or 7.62-mm machine gun	N/A[d]

Table 2.6—Continued

Type	Country of Origin	Weight (Tons)	Armament	Max Armor (mm)
AML 90 armored car	France	6.00	90-mm gun; 7.62-mm machine gun	12
VCR APC	France	8.70	12.7-mm machine gun or ATGM	12
M3 APC	France	6.70	7.62-mm machine gun	12

SOURCES: Foss, *Jane's Tanks and Combat Vehicles Recognition Guide*, pp. 66–77, 138–39, 168–179, 250–253, 346–349, 416–417; Frank N. Schubert and Theresa L. Kraus, eds., *The Whirlwind War*, Washington, D.C.: Center of Military History, U.S. Army, 1995, pp. 271–275. See Schubert and Kraus, *The Whirlwind War*, p. 271, for a description of Iraqi armor:

> The Iraqis employed items captured from Iran and Kuwait as well as those purchased on the international arms market. Their practice of battlefield reclamation, together with their upgrades and modifications, produced an assortment of unique equipment made from mix-and-match parts.

[a] Classified (composite or steel).

[b] Classified.

[c] 14mm on ATGM versions.

[d] Classified (steel, two layers).

forces initiated the air-campaign phase of Operation Desert Storm on January 17, 1991. By this time, coalition forces were largely in the theater and the process of positioning ground maneuver and support units for offensive operations was in full swing. Task Force Shepherd, in addition to its covering operations, supported combined-arms raids against Iraqi positions on the border.[110] As part of this preparatory phase for the coming ground offensive, U.S. Marines established a logistical base at Kibrit, some 30 mi from the Kuwait border.[111]

---

[110] Cureton, *United States Marine Corps in the Persian Gulf*, pp. 26–28; Anthony A. Winicki, "The Marine Combined Arms Raid," *Marine Corps Gazette*, Vol. 75, No. 12, December 1991, pp. 54–55.

[111] John F. Newell, III, "Airpower and the Battle of Khafji: Setting the Record Straight," thesis, Maxwell-Gunter Air Force Base, Montgomery, Ala.: School of Advanced Airpower Studies, Air University, 1998, p. 11.

During the evening of January 29, Iraqi forces conducted what would be their only significant offensive during Desert Storm when they launched an attack on three axes with the object of "breaking through the observation post line, seizing Kibrit and Khafji, then driving south to take Mishab."[112] The observation post (OP) line consisted of eight consecutively numbered Saudi police stations that had been occupied by U.S. Marines. The OPs were emplaced in a sand berm that extended along the entire Kuwait–Saudi Arabia border. The sand berm was a formidable obstacle:

> Often as high as 15 feet, the wall of sand blocked all vehicle movement across it and intentionally channelized traffic toward the police stations. That allowed both [U.S. Marine] division commanders to concentrate their screening and reconnaissance forces at key locations rather than dispersing them along the entire border.[113]

The 1st Marine Division was responsible for Posts 4, 5, and 6. Of these, OP 4 was the most important because it straddled the approach to Kibrit. The 2nd Platoon, Company A, 1st Reconnaissance Battalion, occupied OP 4. The four companies of Task Force Shepherd were located behind the OPs, with Company D, 3rd Light Armored Infantry Battalion, placed directly behind OP 4. At this point, the mission of Task Force Shepherd, which had four light armored infantry companies in its task organization, was to provide early warning for the division and to assist in deception operations in support of the coming ground campaign.[114]

At 8:30 p.m. on January 29, Marine spotters at OP 4 detected approximately 35 Iraqi armored vehicles closing on their position.

---

[112] Cureton, *United States Marine Corps in the Persian Gulf*, p. 28.

[113] Cureton, *United States Marine Corps in the Persian Gulf*, pp. 29–30.

[114] Cureton, *United States Marine Corps in the Persian Gulf*, pp. 28–32. Captain Roger L. Pollard commanded Company D, 3rd Light Armored Infantry Battalion, and had arrived in Saudi Arabia in August 1990 with the 7th MEB. His company had 13 LAV-25s, seven LAV-ATs, and one LAV-C2. The LAV-ATs came from an attached antitank section, which Lieutenant Colonel Myers had provided to the light armored companies.

Shortly thereafter, the OP came under Iraqi attack and the battle for OP 4 was joined. The Iraqi formation was a task force consisting of a T-62 tank battalion reinforced by BMP-2s and BTR-60s.[115] Throughout the night and into the next morning, the Marines at OP 4 and elements of Task Force Shepherd fought a battle against Iraqi forces that took place in three phases:

> The reconnaissance platoon came under attack and withdrew. Company D moved to cover the platoon's withdrawal and attempted to delay or halt the Iraqi advance, took casualties and withdrew. Companies A and B from Task Force Shepherd replaced Company D and drove out Iraqis from OP 4 which concluded the engagement.[116]

The U.S. Marines used a combination of ground fire, artillery, and air support to blunt and turn back the attack at OP 4. Additionally, LAV-ATs, whose gunners fired Tube-Launched, Optically Tracked, Wide-Guided (TOW) antitank missiles at long range using thermal sights, proved very effective against Iraqi armor. The LAV-25s, with their 25-mm chain guns, also proved quite potent against Iraqi APCs. The U.S. Marines also innovated during the battle: Because the LAV-25s did not have thermal sights, the U.S. Marines used the thermal sights of LAV-ATs to direct the 25-mm fire of the LAV-25s.[117] Nevertheless, the battle at OP 4 was fundamentally a combined-arms fight. The outnumbered Marines used artillery and air support—provided by AH-1 Sea Cobra attack helicopters, C-130 gunships, AV-8B Harriers, F-16s,F-15Es, and A-10s—to defeat the Iraqi attack. At the end of the battle, 22 abandoned Iraqi tanks and other vehicles littered the area around OP 4.[118] Task Force Shepherd lost 11 personnel and two

---

[115] Newell, "Airpower and the Battle of Khafji," pp. 30, 34–36. See also Cureton, *United States Marine Corps in the Persian Gulf,* pp. 33–41, which indicates the Iraqis employed both T-62 and T-55 tanks against OP 4.

[116] Cureton, *United States Marine Corps in the Persian Gulf,* p. 33.

[117] Cureton, *United States Marine Corps in the Persian Gulf,* pp. 33–40.

[118] Newell, "Airpower and the Battle of Khafji," p. 37.

vehicles to friendly fire.[119] In the aftermath of the battle, Task Force Shepherd continued its screening and reconnaissance missions in preparation for the ground-attack phase of Desert Storm.

Operation Desert Storm. On February 21, three days before the ground-attack phase of Desert Storm commenced, all elements of the 1st Marine Division were in their attack positions behind the sand berm. In the early morning hours of February 24, G-Day, the 1st Marine Division breached the berm and Iraqi obstacle belts and began its attack into Kuwait. Task Force Shepherd moved through the obstacle belts and set up a screen to cover the division's attack on the Al Jaber airfield. There they encountered a "surrealistic battlefield where visibility at 1500 [hours] was down to 50 or 100 meters." This poor visibility was caused by the smoke from hundreds of burning oil wells that the Iraqis had set on fire. Task Force Shepherd engaged in a number of minor skirmishes as it established the screen. The greatest impediment to the force's movement was the mobs of Iraqis trying to surrender. Task Force Shepherd's historian, Captain John F. McElroy, later recalled that "[l]iterally thousands of Iraqis emerged, at times, begging for food."[120]

In the early morning hours of February 25, the Iraqis counterattacked. They were engaged by Company D, Task Force Shepherd, along the screen line. Another Iraqi attack was met by Company C, Task Force Shepherd, which was screening the division command post. The Iraqi attacks were defeated. On February 26, the division resumed its attack toward Kuwait City, pursuing its objectives of securing the Kuwait International Airport and blocking the exits out of Kuwait City.[121] Task Force Shepherd's mission was "to skirt the east side of the airport and seize the highway system to the northeast and secure the division's right flank and isolate the airport from the east."[122]

---

[119] Hofmann and Starry, eds., *From Camp Colt to Desert Storm*, p. 487.

[120] Cureton, *United States Marine Corps in the Persian Gulf*, pp. 50–96, quotations from p. 83.

[121] Cureton, *United States Marine Corps in the Persian Gulf*, pp. 90–109; Melon et al., *Anthology and Annotated Bibliography*, pp. 141–142.

[122] Cureton, *United States Marine Corps in the Persian Gulf*, p. 109.

By 6:30 p.m. on February 26, Task Force Shepherd had reached its screening positions. At 10:30 p.m., General Myatt ordered Task Force Shepherd to replace Task Force Taro as the division task force responsible for occupying the airport. (Task Force Taro was delayed in the Al Burquan oil field.) Task Force Shepherd began the attack on the airport at 4:30 a.m. on February 27 and quickly routed the Iraqi forces in the airport complex. In a few hours, Task Force Taro arrived and the airport was quickly secured. At 9:00 a.m., General Myatt arrived at the airport with the division forward-command post. Having secured its objectives, the division ceased offensive operations at 6:47 a.m. on February 28. The division remained in position at the Kuwait International Airport until March 5. The withdrawal of divisional units back to Saudi Arabia, the first leg of the journey home, was ordered on March 1. Task Force Shepherd was the first to depart.[123]

General Hopkins believes that Marine LAVs made a significant contribution to U.S. Marine operations in Desert Shield and Desert Storm:

> The 25-mm chain gun was deadly. The LAV held up. It could go 30 to 40 miles per hour across the desert floor. We used it when we were determining where we were going to breach and before G-Day, we used the LAV to run up and down the border of Kuwait to confuse the Iraqis on where our penetration was going.[124]

Additionally, the LAV-AT with its TOW ATGMs and thermal sights proved particularly effective against Iraqi armor. The other LAV variants did not have this capability, however, and

> [p]assive night sights were inadequate because they required more ambient illumination than was always available and because they provided no day or night capability on an obscured battlefield.

---

[123]Cureton, *United States Marine Corps in the Persian Gulf*, pp. 110–121.

[124]Melon et al., *Anthology and Annotated Bibliography*, p. 32. After the 1st Marine Division was in place, General Hopkins left command of the 7th MEB to assume duties as the deputy commander of I Marine Expeditionary Force.

> The LAV-AT, however, was equipped with AN/TAS-4 thermal sights for its TOW missiles, which the Marine Corps called "the single most significant system enhancement of Operation Desert Shield/Desert Storm." . . . Without thermal imaging, the LAV battalion experienced severe operational restrictions in low visibility conditions. As a result, it was recommended that the LAV be equipped with thermal sights and vision devices.[125]

Key Insights

Task Force Shepherd's experiences during the Gulf War offer the following overarching insights that highlight the important role that medium-armored forces can play in modern combat—particularly when they are employed as integral components of a combined arms force:

- The LAV provided Task Force Shepherd with protected mobility compared to U.S. Marine light infantry, who were transported in AAV-7s or cargo trucks.
- This medium-armored force provided the 1st Marine Division with a highly mobile task force that could screen the division and rapidly reposition itself (compared to U.S. Marine tank units) in response to changing orders.
- The ability of Task Force Shepherd units to execute combined-arms fights that integrated artillery and air support was a crucial leveler in situations where those units faced numerically superior Iraqi forces (e.g., at OP 4). The rapid availability of fires was also crucial in several instances.

---

[125] Cordesman and Wagner, *The Gulf War*, p. 705.

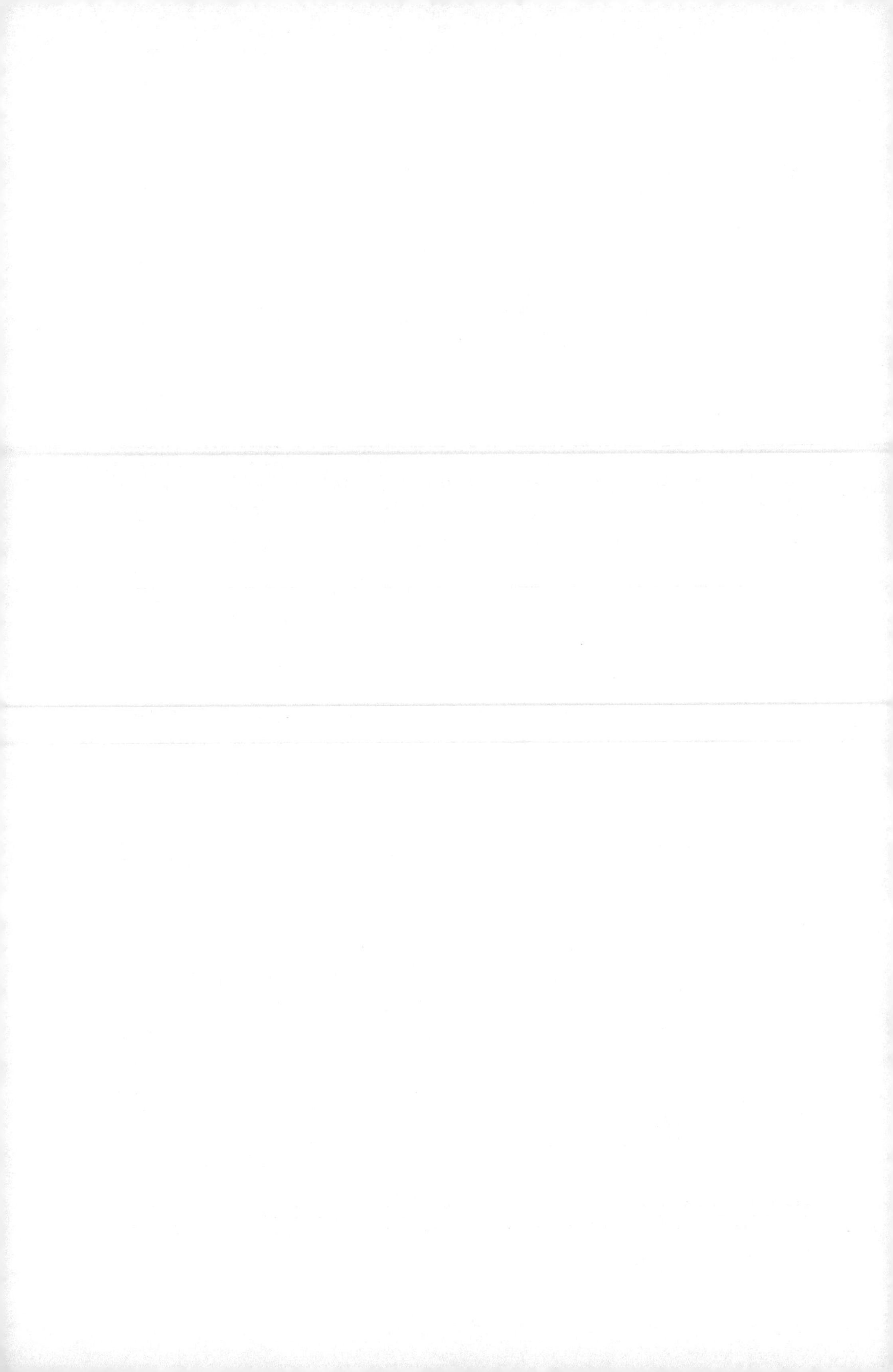

CHAPTER THREE

Medium-Armored Forces in the Center of the Range of Military Operations

This chapter assesses six cases of the use of medium-armored vehicles in operations that fall below in the center of the range of military operations. They include

- the Soviet strike operation in Czechoslovakia (1968)
- South Africa in limited conventional conflict and counterinsurgency operations in Angola (1975–1988)
- Soviet strike and counterinsurgency operations in Afghanistan (1979–1989)
- the U.S. strike operation in Panama (1989)
- the Russian Federation in limited conventional conflict and counterinsurgency operations in Chechnya (1994–2001)
- SBCTs in counterterrorism and counterinsurgency operations in Iraq (2003–2005).[1]

Soviet Airborne Operations in Czechoslovakia (1968)

In January 1968, the Czechoslovakian reformer Alexander Dubcek replaced Antonin Novotny as First Secretary of the Czechoslovakian Communist Party. Secretary Dubchek promised his people "socialism with a human face." In April the Czechoslovakian Communist

[1] This case analyzes this two-year period only. As of this writing, SBCTs remain in Iraq.

Party published an "Action Program" that called for extensive political and economic reforms.[2] That same month, the Soviet General Staff began planning for a military intervention, and in the coming months it conducted exercises and a surreptitious mobilization to prepare for an invasion of Czechoslovakia.[3]

In May the Soviet leadership made its concerns about the developing situation clear to the Czechs. Nevertheless, the Czechs pressed forward and abolished censorship in June.[4] On July 15, representatives of the Communist parties of the Soviet Union, Hungary, Poland, East Germany, and Bulgaria met in Warsaw. They warned the Czechs that "the situation in Czechoslovakia jeopardize[d] the common vital interests of other socialist countries."[5] The Soviets also began military preparations in the event that intervention was deemed necessary.

The situation continued to deteriorate. On August 3, the Warsaw Pact leadership met in Bratislava. At that meeting, Leonid Brezhnev, General Secretary of the Communist Party of the Soviet Union, announced the Brezhnev Doctrine of limited sovereignty in a seemingly reconciliatory move.[6] The limits of the doctrine were, however, clear: Sovereignty was "to be exercised only as long as it did not damage the interests of the 'socialist commonwealth' as a whole."[7] At the same meeting, Secretary Brezhnev received a letter from members of the Czechoslovakian Presidium, warning that "the socialist order is under threat" and requesting a military intervention in Czechoslova-

---

[2] Matthew Frost, "Czech Republic: A Chronology of Events Leading to the 1968 Invasion," RadioFreeEurope/RadioLiberty, August 20, 1998.

[3] Steven J. Zaloga, *Inside the Blue Berets: A Combat History of Soviet and Russian Airborne Forces, 1930–1995*, Novato, Calif.: Presidio Press, 1995, p. 157.

[4] Frost, "Czech Republic."

[5] Frost, "Czech Republic."

[6] Frost, "Czech Republic."

[7] Raymond E. Zickel, *Soviet Union: A Country Study*, Washington, D.C.: Library of Congress, 1989, Appendix C.

kia.[8] At an August 18 secret meeting in Moscow, the Warsaw Pact leaders (less Czechoslovakia) authorized the invasion of Czechoslovakia.[9]

The repression of the Hungarian uprising in 1956 had been the last occasion when the Warsaw Pact had intervened militarily in the internal affairs of a member. In that intervention, 200,000 Soviet troops crushed the rebellion in five days, leaving 25,000 Hungarians dead.[10] The Soviets suffered 669 troops killed, 51 missing, and 1,540 wounded.[11] The Czechoslovakia of 1968, however, differed in one important aspect from the Hungary of 1956:

> The Hungarian Army in 1956 was small and poorly developed; its officers were cowed by political purges and secret police infiltration. In contrast, the Czechoslovak army in 1968 was one of the most modern in the Warsaw Pact, well trained and well equipped. . . . [T]here was always the risk that Dubcek's alluring promises of reform would subvert their allegiance from the Warsaw Pact to Czechoslovak nationalism. If the Czechoslovak Peoples Army (CSLA) decided to fight, an invasion would quickly become far bloodier than the 1956 Hungarian uprising.[12]

Therefore, any military intervention had to be swift and focused on neutralizing the CSLA.

The plan developed for the invasion of Czechoslovakia, code-named Operation Danube, involved an airborne and land invasion. Rapid decapitation of the government lay at the heart of the plan because "the Soviets presumed that without official orders the Czechoslovak Army would stay in its barracks."[13] To execute this rapid strike,

---

8 Zickel, *Soviet Union*, Appendix C.

9 Zaloga, *Inside the Blue Berets*, p. 160.

10 Zickel, *Soviet Union*, Appendix C.

11 Lester W. Grau and Mohammand Yahya, "The Soviet Experience in Afghanistan," *Military Review*, Vol. 75, No. 5, October 1995, p. 18.

12 Zaloga, *Inside the Blue Berets*, p. 157.

13 Zaloga, *Inside the Blue Berets*, p. 157.

the Soviets relied upon their *Vozdushno-Desantniy Voisk* (VDV) Airborne Force.

This section assesses the role of the VDV in the critical mission of decapitating the Czechoslovakian government.

The Armored Forces

The VDV forces of the Soviet Army possessed several types of medium-armored vehicles. The principal vehicles employed during Operation Danube were the ASU-57 and ASU-85 tracked assault vehicles and the BRDM-2 wheeled armored vehicles. Table 3.1 provides the characteristics of these vehicles.

Employment

The Soviet operational plan for the invasion of Czechoslovakia called for VDV parachute and air-landing operations at multiple locations within Czechoslovakia and cross-border attacks by Warsaw Pact forces. The 7th Guards Airborne Division was tasked with the critical mission of decapitating the Czechoslovakian government in Prague.[14]

The invasion began on the evening of August 20. At approximately 8:30 p.m., a Soviet Air Force air-control aircraft, in deceptive *Aeroflot* markings, landed at Prague's Ruzyne airport. This airplane and its crew were on hand to provide air-traffic control for the arrival of the 7th Guards Airborne Division in the event that the airport's own control tower was knocked out during the coming military action. Later that evening, a special *Aeroflot* flight brought in Soviet government officials and *Komityet Gosudarstvennoy Bezopasnosti* (KGB) [Soviet Committee for State Security] agents. At midnight, the Ruzyne airport was closed to civilian traffic and a KGB special operations team arrived.[15]

At 3:37 a.m., under the cover of MiG-21 fighters, the 7th Guards Airborne Division started air-landing at the Ruzyne airport. In short order, one of its assault companies sealed off the airport while the KGB special operations team took over the airport's control tower. With the airport secured, further elements of the 7th Guards Airborne Division

---

[14] Zaloga, *Inside the Blue Berets*, p. 158.

[15] Zaloga, *Inside the Blue Berets*, pp. 160–161.

Table 3.1
Soviet Airborne (VDV) Armored Vehicles in Czechoslovakia

Type	Weight (Tons)	Armament	Max Armor (mm)
ASU-57 assault gun	3.7	57-mm gun	6
ASU-85 assault gun	15.4	85-mm gun; 7.62-mm machine gun	40
BRDM-2 amphibious reconnaissance vehicle	7.7	14.5-mm machine gun; 7.62-mm machine gun	10
BRDM-2 antitank version	7.7	ATGM	14

SOURCES: Foss, *Jane's World Armoured Fighting Vehicles*, pp. 168–170, 331–34; Foss, *Jane's Tanks and Combat Vehicles Recognition Guide*, pp. 286–89.

began arriving every 30 seconds. By 4:30 a.m., special assault groups, composed of elements of the division's reconnaissance company and the 108th Guards Paratrooper Regiment, had taken two objectives. These two assault groups had "a few ASU-85 and ASU-57 assault guns, plus a small number of BRDM-2 antitank vehicles."[16] They also commandeered two buses and four cargo trucks. One column took the presidential palace in Prague and arrested Ludwig Svoboda, President of Czechoslovakia. The other column of paratroopers, led by KGB personnel, took the Czechoslovakian Communist Party headquarters in Prague. There, they arrested Secretary Dubcek and some two dozen other top government officials. Although these officials had been alerted to the invasion, "they presumed it would take ten to eleven hours for the Soviet Army to drive into Prague from the Polish or German border. The paratroopers' sudden appearance completely surprised them."[17]

By 7:00 a.m. on August 21, the mission of decapitating the Czechoslovakian government was accomplished. As other elements

[16] Zaloga, *Inside the Blue Berets*, p. 161.

[17] Zaloga, *Inside the Blue Berets*, p. 161; see also pp. 158 and 163. Another Soviet airborne force, the 103rd Guards Airborne Division, had the mission of taking key Czechoslovakian Army headquarters, which they accomplished. Consequently, the Czech General Staff never had the opportunity to issue instructions about resistance. These actions, coupled with the decapitation of the government, ensured that the Czechoslovakian Army was neutralized.

of the 7th Guards Airborne division arrived, they were deployed to key locations within Prague, including government buildings, communication centers, and power plants.[18] One of the final objectives to fall to the paratroopers was the capture of the main radio station on Vinhradska Street. Czech civilians had barricaded the street and were reinforcing it when the paratroopers arrived in ASU-85s and *Boyevaya Razvedyuatel'naya Dozornaya Meshinas* (BRDMs) [Russian combat reconnaissance patrol vehicles]. Before assaulting the radio station, the soldiers used these vehicles for protection from the crowd and to destroy the barricade.[19]

Later on August 21, tank and motor rifle units began arriving in Prague from the Central Front, and Warsaw Pact forces gained complete control of the city. Nevertheless, it was rapid insertion of the VDV forces that had enabled the coup de main. Furthermore, their "sudden appearance had prevented Czech civilians from seizing weapons from police stations or army barracks, as had occurred in Hungary in 1956."[20] The Soviets suffered only 96 killed in action.[21] Casualties among Czechoslovakian citizens in "the first week of the main conflict . . . [totalled] something like more than 70 people killed and several hundred wounded."[22]

Key Insights

Soviet operations in Czechoslovakia offer insights about the utility of rapidly deployable medium-armor units, although it is important to note the absence of determined Czech resistance to these forces. The following key insights emerge:

---

[18] Kenneth Allard, "Soviet Airborne Forces and Preemptive Power Projection," *Parameters*, Vol. 10, No. 4, December 1980, p. 43.

[19] Zaloga, *Inside the Blue Berets*, pp. 162–163.

[20] Zaloga, *Inside the Blue Berets*, p. 163.

[21] Grau and Yahya, "The Soviet Experience in Afghanistan," p. 18.

[22] Brian Kenety, "MPs Agree on Compensation for Victims of 1968 Soviet-Led Invasion," *Czech Radio*, February 25, 2002.

- The operations of the 7th Guards Airborne Division in Operation Danube were a textbook example of a coup de main. They showed the value of well-trained, rapidly deployable forces.
- The medium-armored vehicles used by the VDV were air-landable and gave the paratroopers protected mobility and sufficient firepower to accomplish their missions. These vehicles also provided an intimidation factor unavailable to truck-mounted infantry. It is doubtful that tank units could have made the rapid dash to the capital that the VDV's medium-armored vehicles accomplished.
- The operation demonstrated the importance of intelligence, cooperation with special operating forces (the KGB, in this case), and surprise. Nevertheless, the operation may have gone quite differently if the Czech Army had not been neutralized and if the citizenry had resisted like the Hungarians during the 1956 Hungarian Revolution.

South Africa in Angola (1975–1988)

In 1974 leftists in Portugal successfully deposed the country's authoritarian dictatorship in the nearly bloodless Carnation Revolution. This leftist coup reverberated throughout the country's residual colonial possessions. In Angola, it precipitated the end of Portuguese administration and led to a fierce, multisided civil war. The primary feature of the Angolan Civil War was the competition for primacy between the Movement for the Popular Liberation of Angola (MPLA) and the National Union for the Total Independence of Angola (UNITA). The MPLA was heavily supported by the Soviet Union and Cuba between 1975 and 1990. UNITA was supported by the Republic of South Africa between 1975 and 1994. A number of other outside sponsors entered and left the fray over time,[23] but MPLA and UNITA were the domi-

[23] Other internal participants included Holden Roberto's National Front for the Liberation of Angola movement and secessionists in the enclave of Cabinda. External participants included a variety of actors such as the armed wing of the African National Congress, Namibian insurgents from the Southwest African People's Organization (SWAPO), the Gulf

nant domestic players. They would contest Angola's fate for more than 25 years.

The military component of the Angolan Civil War featured lengthy periods of desultory, semiconventional skirmishing punctuated by occasional, large-scale conventional campaigns. Over the course of the war, MPLA generally held the initiative on the conventional battlefield. During the typical campaign season, MPLA would launch a small number of multibrigade offensives against UNITA strongholds in southeastern Angola.[24] Despite tremendous materiel and advisory support from the Soviet Union and Cuba, MPLA offensives rarely achieved any lasting effect. Generally speaking, UNITA was able to repulse MPLA columns before they threatened key political and military objectives. Covert materiel and advisory support from South Africa and the United States often played a significant role in UNITA's ability to repulse MPLA offensives.[25]

UNITA occasionally launched conventional offensives against MPLA–occupied territories when its relative strength was on the upswing. However, MPLA grew more effective over time as increasing amounts of sophisticated Soviet equipment flowed into MPLA's stocks and as the Soviet and Cuban presence expanded. (Cuba's presence eventually exceeded more than 30,000 combat troops.[26]) For most of

---

Oil Company, the De Beers mining conglomerate, the British Lohnro Corporation, and the governments of Britain, China, Egypt, France, the Gulf States, Saudi Arabia, South Africa, the United States, Zaire, and Zambia, to name only the most prominent. For the political history of this conflict, see Chester Crocker, *High Noon in Southern Africa: Making Peace in a Rough Neighborhood*, New York: W.W. Norton, 1992; Robert Jaster, *The 1988 Peace Accords and the Future of South-Western Africa*, Adelphi Paper 253, London: International Institute for Strategic Studies, 1990; and William Minter, *King Solomon's Mines Revisited*, New York: Basic Books, 1986.

[24] Jaster, *The 1988 Peace Accords*, pp. 9–25.

[25] Peter Stiff, *The Silent War: South African Recce Operations, 1969–1994*, Cape Town: Galago Publishers, 1996, pp. 182–216, 234–239, 351–368.

[26] Other estimates for the Cuban contingent range as high as 50,000. By 1985, there were confirmed to be 30,000 Cubans and over 3,000 Soviet and East German advisors. These forces were further reinforced in 1987–1988. See Crocker, *High Noon in Southern Africa*, pp. 208, 355–356.

the conflict, UNITA's ability to take the war into MPLA territory was largely limited to unconventional operations.

Aside from its efforts to eliminate UNITA, MPLA also provided staging and support facilities to guerillas operating inside South African territory. Several battalions of Mkonto we Sizwe, the armed wing of Nelson Mandela's African National Congress (ANC), and virtually the entire armed strength of SWAPO were based in Angola. At times, ANC and SWAPO forces were called upon to assist MPLA operations against UNITA.[27]

South Africa supported UNITA as an anticommunist alternative to MPLA and to prevent SWAPO and ANC guerrillas from establishing bases in southeast Angola (from which the guerillas could more easily operate against South African territory).[28] South African Army (SAA) conventional operations in Angola followed two models. The first involved South African mechanized and airborne or airmobile raids against SWAPO base areas up to 250 km deep inside Angola, at times requiring collateral contact with MPLA forces. The second involved direct South African action against the occasional MPLA offensive that appeared capable of overwhelming UNITA and ousting it from its sanctuary in southeastern Angola.[29] Overall, the SAA conducted 16 major conventional operations in Angola between 1976 and 1988.[30]

The Armored Forces

The armored vehicles used by the SAA are shown in Table 3.2. The standard SAA IFV of this period was the Ratel-20. Armed with a tur-

---

[27] Discussion on July 26, 2002, between Adam Grissom and a former Mkonto we Sizwe officer deployed in Angola during the conflict. The officer wished to remain anonymous.

[28] Willem Steenkamp, *Borderstrike! South Africa into Angola*, Durban, South Africa: Butterworth, 1983, p. 16.

[29] Helmoed-Romer Heitman, "Operations Moduler and Hooper," in A. de la Rey, ed., *South African Defence Review*, Durban, South Africa: Walker Ramus, pp. 275–294.

[30] These included operations Savannah (1975), Reindeer (1978), Safraan (1978), Restok (1979), Sceptic (1980), Klipklop (1980), Carnation (1981), Protea (1981), Super (1982), Meebos (1982), Phoenix (1983), Askari (1983), Moduler (1987), Packer (1988), and Displace (1988).

Table 3.2
South African Armored Vehicles in the Angolan Border War

Type	Weight (tons)	Armament	Max Armor (mm)
Eland 60 light armored car	6.00	60-mm mortar; two 7.62-mm machine guns	12
Eland 90 light armored car	6.00	90-mm gun; two 7.62-mm machine guns	12
Ratel-20 IFV	20.35	20-mm cannon; three 7.62-mm machine guns	20
Casspir APC	13.80	up to three 7.62-mm machine guns	N/A[a]
Buffel APC	6.50	7.62-mm machine gun	N/A[b]

SOURCES: Foss, *Jane's Tanks and Combat Vehicles Recognition Guide*, pp. 252–253, 296–297, 380–381; "Armoured Personnel Carries (Wheeled), South Africa," *Jane's Armour and Artillery 2001–2002*, n.d.

[a] Classified.

[b] Protection from small arms and mines.

ret-mounted 20-mm cannon and a coaxial machine gun, the Ratel-20 was operated by three crewmen and carried nine infantry soldiers in its rear compartment.[31] SAA units also used Buffel and Casspir APCs, which, though lightly armed and armored, provided excellent protection for their crews and embarked infantry sections against the ubiquitous land-mine threat in Angola. The SAA also utilized the Eland armored car for fire support and antitank duties in Angola. The Eland, a locally produced variant of the French Panhard family of vehicles, is armed with a low-pressure, 90-mm main gun or a 60-mm mortar. Each variant also includes machine guns.[32]

The SAA favored wheeled medium-armored vehicles, like the Ratel, Eland, Casspir, and Buffel, because of their operational radius and reliability. Nearly all South African support vehicles used during

---

[31] Helmoed-Romer Heitman, *War in Angola: The Final South African Phase*, Gibraltar: Ashanti, 1990, p. 348.

[32] Heitman, *War in Angola*, p. 348.

the war were also wheeled, including the G-6 self-propelled artillery system. The limited exception was the SAA's Olifant MBT. The Olifants, an upgraded variant of the British Chieftain MBT, were deployed to Angola in small numbers in the last two years of the war. However, their role was quite minor because the SAA relied heavily on wheeled medium-armored vehicles.

The armored vehicles employed by MPLA are shown in Table 3.3. MPLA forces fielded T-62, T-54/55, and T-34 MBTs, BRDM and PT-76 reconnaissance vehicles, and BMP-1 and BTR-60 APCs.[33] The MBTs employed by the MPLA forces had much heavier armor, more-lethal main armaments, better antitank ammunition, and better fire-control systems than the Eland and other assorted vehicles pressed into antitank duty by the SAA and UNITA.[34] The Angolan BMPs also had more lethal main armaments (73-mm gun and antitank guided missiles) and thicker armor than the equivalent South African vehicles, though the BTR was more comparable. Additionally, MPLA brigades used a great number of tracked and wheeled surface-to-air systems (e.g., SA-8, ZSU-23-4), engineer vehicles (i.e., mobile bridging units), command vehicles, and logistics vehicles. SAA units committed to Angola lacked many of these heavy support vehicles.

South Africa employed medium-armored vehicles in Angola, while MPLA had MBTs. The South African armored formations used in Angola were also medium-armored in comparison to the heavy Olifant MBT available to the SAA.

Employment

South African forces were typically employed in task-organized "combat groups" built around a mechanized or motorized infantry battalion. The battalions most often used as the core of task forces were

---

[33] See Fred Bridgland, *The War for Africa: Twelve Months that Transformed a Continent*, Gibraltar: Ashanti Publishers, 1991, p. 78. By 1985, MPLA was credited with fielding about 500 MBTs and thousands of other Soviet armored vehicles.

[34] Bridgland, *The War for Africa*, p. 13. In addition to the Eland, the SAA and UNITA utilized a hodgepodge of recoilless rifles, light rockets, antitank guided missiles, and other systems for the direct-fire antiarmor role. However, the Elands were the primary antitank assets available to SAA units.

Table 3.3
Angolan Armored Vehicles in the Angolan Border War

Type	Country of Origin	Weight (tons)	Armament	Max Armor (mm)
T-34/85	Soviet Union	35.20	85-mm gun; two 7.62-mm machine guns	90
T-54/55 tank	Soviet Union	39.00	100-mm gun; 12.7-mm machine gun; two 7.62-mm machine guns	203
T-62 tank	Soviet Union	44.00	115-mm gun; 12.7-mm machine gun; 7.62-mm machine gun	242
PT-76 reconnaissance vehicle	Soviet Union	16.06	76.2-mm gun; 7.62-mm machine gun	14
BMP-1 IFV	Soviet Union	14.80	73-mm gun; 7.62-mm machine gun; ATGM	33
BRDM-2 amphibious reconnaissance vehicle	Soviet Union	7.20	14.5-mm machine gun; 7.62-mm machine gun	10
BTR-60 APC	Soviet Union	11.40	14.5-mm machine gun; 7.62-mm machine gun	9

SOURCES: Foss, *Jane's Tanks and Combat Vehicles Recognition Guide*, pp. 70–77, 178–179, 286–287, 416–417; and Foss, *Jane's World Armoured Fighting Vehicles*, pp. 77–79.

the 32nd "Buffalo" Battalion, 61st Mechanized Battalion, 2nd South African Infantry Battalion, 4th South African Infantry Battalion, and 101st Southwest African Task Force.

The SAA organized its task forces as combined-arms formations, integrating several companies of motorized infantry, one or more squadrons (companies) of antitank Elands, a battery of mortars, a battery of multiple-launch rocket vehicles, a battery of tube artillery (the G-5 or G-6), a small number of "recce" special forces teams, and a variety of support elements.[35] In addition to standard maintenance and supply formations, these support elements included specialized elements such

---

[35] Bridgland, *The War for Africa*, pp. 76, 84, 92.

as electronic warfare teams, South African Air Force (SAAF) liaison teams, and liaison officers for coordinating operations with UNITA units.[36] SAA combat groups deployed to Angola therefore combined a wide variety of capabilities into a relatively small package.

SAA task forces were employed in a limited fashion. They were committed to Angola for particular operations with specified time-frames and limited objectives. South African forces generally concluded their operations by withdrawing from Angolan soil rather than remaining in sustained contact with MPLA and Cuban forces.

South African operations in Angola tended to be quite dispersed. Tens or hundreds of kilometers might separate individual task forces, and their constituent elements were likewise dispersed across massive areas. Operations by all sides gravitated around major airheads, which served as the primary staging bases for supporting dispersed formations.

The specific examples of operations Moduler and Reindeer illuminate general SAA operations. The South Africans embarked upon Operation Moduler in 1987 when signals intelligence (SIGINT) revealed that MPLA had planned a major offensive intended to capture an important UNITA supply center at Mavinga.[37] As planned by Soviet General Konstantin Shaganovitch, the MPLA offensive would involve eight MPLA mechanized brigades and use the MPLA base at Cuito Canavale as its jumping off point.[38] Both Mavinga and Cuito Canavale derived their importance from their large all-weather airfields. South African and UNITA officials agreed that UNITA would be unable to withstand an offensive on this scale. Three SAA battalion task forces were deployed to counter the MPLA offensive. These task forces were widely dispersed and operated in close coordination with UNITA units against those specific MPLA brigades that posed the greatest threat to Mavinga. During the weeks of brutal, close-quarter fighting that ensued, the MPLA offensive was halted and MPLA forces were severely beaten. In Operation Moduler, the South Africans suf-

---

[36] Bridgland, *The War for Africa*, pp. 76, 84, 92.

[37] Heitman, "Operations Moduler and Hooper."

[38] Heitman, "Operations Moduler and Hooper."

fered 43 killed and 90 wounded, plus the loss of seven armored vehicles. UNITA losses were some 270 killed and an unknown number wounded. MPLA losses were estimated at 7,000 killed and an unknown number wounded. MPLA lost 94 tanks, 158 armored personnel carriers, 59 other armored vehicles, 51 artillery pieces, and 377 logistics vehicles.[39]

Operation Reindeer is a useful comparison because South African forces battled SWAPO guerrilla bases in Angola rather than conventional MPLA forces. The operation involved two major components. Elements of the 44th South African Airborne Brigade initiated the operation by airdropping into a major SWAPO base area nearly 250 km inside Angola. With substantial close air support from SAAF Canberra, Impala, and Mirage aircraft, the paratroopers then attacked the base complex. SAAF helicopter units subsequently lifted the paratroopers out. At the same time, a SAA mechanized task force struck across the border against a number of SWAPO bases that were supporting cross-border guerrilla operations. Together, the elements of Operation Reindeer are credited (by South African sources) with killing nearly 600 SWAPO guerrillas and wounding 340 more, at a cost of four South African dead and an unknown number wounded.[40]

MPLA and SAA armor doctrine were a study in contrasts. MPLA operations featured massive, multibrigade sweeps by mechanized infantry and armor formations. MPLA brigades tended to move as complete entities, and often bunched together on the available lines of communications (LOCs). This was due to a mix of logistics constraints and a doctrine that kept MPLA ground forces within the surface-to-air missile umbrella (even after the SAAF largely ceded air superiority in the late 1980s to Cuban- and MPLA-piloted MiG-21s, MiG-23s, and SU-22s).[41]

---

[39] Heitman, "Operations Moduler and Hooper."

[40] Morgan Norval, *Death in the Desert: The Namibian Tragedy*, Washington, D.C.: Selous Foundation Press, 1989, p. 140 (of the online version).

[41] Heitman, "Operations Moduler and Hooper."

SAA medium-armor units were employed in much smaller, more dispersed packages.[42] In contrast to MPLA's broad offensives, South African offensive operations took the form of deep-penetration, blitzkrieg-type operations.[43] For example, South Africa opened its conventional involvement in Angola with Operation Savannah, which featured offensive operations conducted by three dispersed mechanized battalion task forces. The task forces penetrated deeply into Angola to induce operational paralysis in MPLA forces and threaten the MPLA government with a coup de main. In a matter of weeks, South African pincers almost reached to the Angolan capital of Luanda—some 3,000 km from the border.[44] There is little doubt that the South Africans would have seized Luanda had the United States not intervened to prevent a major superpower crisis.

The use of medium armor also enabled the SAA to move its forces by transport aircraft. The wheeled Elands, Ratels, Casspirs, and Buffels were often deployed to the theater, and redeployed within it, by C-130 Hercules and C-160 Transall airlifters. The dependability of the wheeled vehicles also allowed SAA units to deploy considerable distances by surface movement. These attributes created useful options for South African commanders and political authorities in Pretoria. Again, Operation Savannah provides an excellent example. South Africa entered the war in Angola by executing a vertical envelopment on MPLA forces attacking UNITA positions in southeast Angola. Operation Savannah required the deployment of squadrons of Elands to the remote South African airbase at Rundu, then further airlifting them more than 700 km over the border to the UNITA base in Nova Lisboa, Angola.[45]

SAA operations also relied heavily on the thorough integration of special forces operators, called "recces." The SAA has a world-renowned

---

[42] Bridgland, *The War for Africa*, pp. 55–70.

[43] Bridgland, *The War for Africa*, pp. 17, 64. Indeed, Soviet advisors planned many MPLA operations. General Konstantin Shaganovitch assumed command of the combined MPLA, Cuban, Soviet, and East German forces in Angola in 1985.

[44] Steenkamp, *Borderstrike!*, p. 3.

[45] Bridgland, *The War for Africa*, p. 8.

special operations capability resident in its 4th Recce Regiment. During the war in Angola, the regiment conducted independent operations in support of conventional South African forces.[46] One famous example is the recce's shadowing of the key MPLA airbase in southern Angola. After the MPLA fielded MiG-23s to the region, the SAAF lost the ability to contest air superiority. South Africa's response was to infiltrate recces to locations very near the airbase; from these locations, the recces reported departures and arrivals. When MiG-23s were launched, South African air operations were shaped to avoid interception. When ground attack aircraft were launched, South African ground operations—such as fire support and counter-battery operations by G-5 batteries—were shifted to avoid attack until the threat had receded. This unique recce mission, which was sustained for some time, allowed SAAF and SAA operations to proceed without effective intervention from MPLA aircraft.[47]

In addition to independent missions, recce teams were integrated into South African battalion task forces.[48] The recces were absolutely vital to South Africa's success on the ground. They were perhaps the most valuable source of intelligence for South African commanders, as they operated deeply and covertly behind MPLA lines. Often establishing their observation posts only meters from MPLA positions, recce teams shadowed MPLA units and continually reported on MPLA strength and location. Additionally, the recces frequently called in bombardments from South African artillery batteries up to 40 km away, causing casualties among MPLA units and eroding their morale.[49]

Another key aspect of SAA operations was the very aggressive use of artillery. The South Africans were fortunate to have, in the G-5, perhaps the finest tube artillery piece in the world.[50] With a range of more

---

[46] Bridgland, *The War for Africa*, p. 40.

[47] Bridgland, *The War for Africa*, p. 67.

[48] R. S. Lord, "Operation Askari: A Sub-Commanders Retrospective View of the Operation," *Military History Journal of the South African Defense Force,* Vol. 22, No. 4, 1992.

[49] Heitman, *War in Angola*, p. 343.

[50] Bridgland, *The War for Africa*, p. 87.

than 40 km and extraordinarily lethal airburst 155-mm ammunition, the G-5 may have caused more MPLA casualties than any other South African weapon.[51] The SAA distributed G-5 batteries to battalion task forces operating in Angola, and supported the G-5 batteries with additional batteries of 81-mm mortars and 127-mm free-flight rockets.[52] Utilizing recce spotters, SAAF liaison teams accompanying friendly ground forces, spotter aircraft, and remotely piloted vehicles (RPVs), used forward-deployed G-5 batteries to pound MPLA positions constantly and to great effect.[53] This bombardment was particularly effective due to the extraordinary accuracy of the G-5, which allowed spotters to target and destroy individual vehicles and fighting positions with very few "walk-in" rounds.[54] Spotter adjustments of a few meters were not uncommon.

In this period, the SAA provided world-class training to all ranks, from enlisted to senior officers. The uncompromising training standards set for all tasks—from tactical drills to campaign planning—provided a (perhaps *the*) decisive advantage over Angolan and Cuban forces.[55] In many cases, this training allowed SAA units to meet and defeat MPLA units equipped with more-lethal and more-survivable vehicles. This was particularly important because the SAA favored mobility over armor, subscribing to the theory that more-mobile forces could decline engagements with heavier and more-lethal adversaries, choosing instead to maneuver into advantageous engagements. However, the close terrain of southern Angola obliged SAA mechanized units to accept very short-range engagements, typically a distance of between 20 and 200 meters.[56] This remained true despite clear South African superiority in what today would be called C4ISR, including the advantages they accrued from SAAF 10 Squadron Seeker RPVs and outstanding sup-

---

[51] Heitman, *War in Angola*, p. 344.

[52] Heitman, *War in Angola*, p. 343.

[53] Bridgland, *The War for Africa*, p. 115.

[54] Bridgland, *The War for Africa*, p. 87.

[55] Heitman, *War in Angola*, p. 342.

[56] Bridgland, *The War for Africa*, p. 80.

port from the recce teams who operated independently and in direct support of the task forces.[57] Survivability of South African mechanized vehicles remained a concern throughout the Border War.

Superior training, air deployabilty, and combined-arms operations compensated for the vulnerabilities of the South African medium-armored forces. Eventually, however, the SAA determined that its medium-armored vehicles were not sufficiently survivable and lethal against MPLA vehicles, particularly late-model Soviet MBTs. Much like U.S. forces in 1944–1945 on the Western Front, the South Africans found that their Ratels required four or five rounds from their 90-mm main armament to disable MPLA MBTs. This forced the Ratels to operate as platoons whenever MBTs were present.[58] Nevertheless, the SAA's superior situational awareness did not allow it to avoid meeting engagements altogether, especially in the close terrain that was characteristic of much of the operational area. As a result, the South Africans expended great energy in fielding the Olifant MBT, armed with the 105-mm L7 main gun. The Olifants, however, arrived too late and in too limited numbers to make a major mark on the campaign.

Key Insights

For much of its protracted war in Angola, South Africa relied on medium-weight armored forces, rapid intratheater mobility, and high levels of situational awareness. These capabilities are also central to evolving U.S. future concepts. Thus, the following insights from this case are particularly useful:

- SAA medium-weight armored units operated at a significant disadvantage in terms of lethality and survivability when compared to heavy MPLA forces.
- Training was key to South African success in Angola. Superior TTP enabled South African mechanized forces to overcome disadvantages in lethality and survivability.

---

[57] Heitman, *War in Angola*, p. 344.

[58] Bridgland, *The War for Africa*, p. 139.

- Task organization was also key to South African success in Angola. It allowed South African forces to operate in a dispersed manner while remaining just small enough to be logistically supportable in a difficult environment.
- Situational awareness was another key to South African success in Angola. South African investments in recce teams, RPVs, spotter aircraft, and electronic warfare allowed small South African forces to overcome their lack of lethality and survivability. The complex terrain (i.e., bush), made meeting engagements unavoidable, however.
- Indirect fires provided most of the SAA's lethality in Angola. Effective indirect fire relied on excellent materiel, extraordinary spotting by recce teams and RPVs, and innovative task organization.
- Despite the training, situational awareness, and combined arms advantages the SAA enjoyed in Angola, the service ultimately decided that those advantages were not sufficient to address the operational environment. The SAA chose to make itself heavier with the introduction of the Olifant MBT.

Soviet Operations in Afghanistan (1979–1989)

On April 27, 1978, the People's Democratic Party of Afghanistan (PDPA) toppled the regime of President Mohammed Daoud in a bloody coup. Subverted elements of Afghanistan's armed forces stormed the presidential palace, killing President Daoud, his brother, and a number of presidential advisers. In the aftermath of the coup, the PDPA created a Revolutionary Military Council, which declared the creation of the Democratic Republic of Afghanistan (DRA). The insurgent army officers also installed Nur Mohammed Taraki, leader of the Khalq faction of the PDPA, as president. Babrak Karmal, head of the PDPA's Parcham faction, became vice-president. On April 30, the Soviet Union recognized the DRA.[59]

---

[59] Anthony H. Cordesman and Abraham R. Wagner, *The Lessons of Modern War:* Vol. III, *The Afghan and Falklands Conflicts*, Boulder, Colo.: Westview Press, 1990, pp. 23–25, 28–29;

The ascension of the PDPA did not, however, bring stability to Afghanistan—quite the contrary. President Taraki's regime, which was of a Stalinist bent, soon "embarked on a series of harebrained schemes to collectivize agriculture and to suppress the role of Islam in Afghan life."[60] These reforms were met with armed resistance in the rural, strongly Islamic areas of Afghanistan. President Taraki turned to the Soviet Union for assistance and on December 5, 1978, concluded the Soviet-Afghan Friendship Treaty that established collective security arrangements between the two nations.

The treaty and the resultant influx of Soviet advisers and military equipment "led many Muslims in the growing Afghan resistance movement to call for a holy war, a jihad, against the Russian infidels they saw lurking behind the Kabul regime."[61] The internal situation steadily worsened. In March 1979 the rebels, or Mujahideen, precipitated an uprising in Herat. The majority of the DRA Army's 17th Infantry Division mutinied and joined the rebels. Eventually, elements of the DRA Army and Air Force units loyal to the government retook the city, at the cost of some 5,000 dead. Among the dead were over 100 Soviets, including Soviet Army advisors, some of whom were tortured and murdered by the Mujahideen.

The Soviet Union, which regarded Afghanistan as a key buffer state, responded to the worsening conditions with more military aid. Soon, Soviet attack-helicopter pilots were supporting DRA Army operations against the Mujahideen. In July a battalion of the Soviet 345th Guards Airborne Regiment deployed to the Bagram air base north of Kabul to secure the Soviet An-12 air-transport regiment located there. As the civil war worsened, large-scale defections from the DRA Army to the Mujahideen increased and the DRA Army shrank from 90,000 to 40,000 soldiers. Furthermore, half of the DRA Army's officers had been purged by the regime or had deserted. The DRA Army was on the verge of disintegration. The number of Soviet advisors in Afghanistan steadily increased, finally reaching 3,000 by September. They

---

Zaloga, *Inside the Blue Berets*, pp. 227–228.

[60] Zaloga, *Inside the Blue Berets*, p. 228.

[61] Zaloga, *Inside the Blue Berets*, p. 228.

attempted to prop up the DRA Army and were directly engaged in fighting the rebellion. Despite their efforts, the rebels scored major victories over DRA Army units in the summer of 1979.[62]

The increasingly widespread civil war was not the only problem facing President Taraki's regime. The PDPA was rife with internal intrigues and power struggles. Indeed, a "diplomatic communiqué from the U.S. embassy tellingly described the PDPA as 'a bottle of angry scorpions all intent on stinging each other'."[63] Since his assumption of power, President Taraki had purged former President Daoud's adherents from power and had also forced Parcham faction followers out of the government and the DPA Army. The political repression of the Taraki regime was further manifested in its murder of some 17,000 Afghans in political purges in the cities. Nevertheless, President Taraki's downfall came at the hands of Hafizullah Amin, a rival in the Khalq, rather than at the hands of outsiders.

Amin's power within the Taraki regime had grown as he gained control of the DRA's internal security forces. In mid-September 1979, Amin began purging President Taraki's ministers from the cabinet. In response, President Taraki unsuccessfully attempted to capture and kill Amin, who escaped. On September 16, Amin ousted President Taraki, later ordering security police to kill the deposed president, which they did by smothering him with a pillow.[64]

The Soviets were alarmed as they "watched this new communist state spin further out of control."[65] A delegation of Soviet generals visited Afghanistan in April 1979. It was led by General of the Army and Main Political Directorate Head A. A. Yepishev, who had led a similar mission to Czechoslovakia before the Soviet invasion in 1968. In August a group of 60 Soviet officers, led by General of the Army I. G. Pavlovski, the commander in chief of the Soviet ground forces, con-

---

[62] Cordesman and Wagner, *The Afghan and Falklands Conflicts*, pp. 24–25, 31; Zaloga, *Inside the Blue Berets*, p. 228; Grau and Yahya, "The Soviet Experience in Afghanistan," p. 18.

[63] Zaloga, *Inside the Blue Berets*, p. 228.

[64] Cordesman and Wagner, *The Afghan and Falklands Conflicts*, p. 31; Zaloga, *Inside the Blue Berets*, pp. 228–229.

[65] Grau and Yahya, "The Soviet Experience in Afghanistan," p. 18.

ducted extensive reconnaissance of Afghanistan, ostensibly "to study the possibility of seizing control of the government and the armed forces." General Pavlovski had commanded the forces that invaded Czechoslovakia in 1968.[66]

In September the Soviets began planning and organizing what they assumed would be a Czechoslovakian-style coup de main against the Amin regime. In the coming months what would become known as the 40th Army began forming for the invasion of Afghanistan. The invasion force consisted of the 103rd Guards Airborne Division, the 345th Separate Parachute Regiment, and the 5th, 108th, and 201st Motorized Rifle Divisions.[67] The Soviets also began infiltrating *Spetsnaz* and other forces into Afghanistan; by December 9, some 2,000 paratroopers from the 103rd Guards Airborne Division had been airlifted to the Bagram air base. The paratroopers dispatched forces to secure the crucial Salang Tunnel.[68]

In the wake of an attempt on President Amin's life by Taraki loyalists, the Soviet Politburo decided that the "turmoil in Kabul was becoming intolerable." Mikhail Suslov,

> the chief idealogue [*sic*] of the Soviet Communist Party, convinced Secretary Brezhnev that a small contingent of Soviet forces should be sent into Afghanistan to install Babrak Karmal of the Parcham faction and restore order.[69]

---

66 Grau and Yahya, "The Soviet Experience in Afghanistan," p. 18; quote from Cordesman and Wagner, *The Afghan and Falklands Conflicts*, p. 30.

67 Edward B. Westermann, "The Limits of Soviet Airpower: The Bear Versus the Mujahideen in Afghanistan, 1979–1989," thesis, Maxwell-Gunter Air Force Base, Montgomery, Ala.: School of Advanced Airpower Studies, Air University, 1997, p. 16.

68 Cordesman and Wagner, *The Afghan and Falklands Conflicts*, p. 33. See also Frank N. Schubert, "U.S. Army Corps of Engineers and Afghanistan's Highways, 1960–1967," *Bridge to the Past*, No. 4, June 1996. The Salang Tunnel, constructed by the Soviet Union in the 1950s, was part of a highway that ran from Kabul across the Hindu Kush Mountains to the Soviet border. The tunnel was 2.4 km long and shortened the traditional route by some 200 miles. Control of the tunnel was a critical objective because it would enable Soviet forces to use the tunnel to enter Afghanistan.

69 Zaloga, *Inside the Blue Berets*, p. 229.

On December 12, the Central Committee of the Soviet Union decided "to provide military assistance to the DRA by means of sending a limited contingent of Soviet forces to its territory."[70]

The Soviet plan envisioned an initial Czechoslovakia-like coup de main, a short period of Soviet presence to stabilize the situation in Afghanistan, and a final rapid withdrawal of Soviet forces. Accordingly, the troops allocated to the 40th Army numbered fewer than 50,000.[71] The Soviet plan focused on six key points:

- Stabilize the country by garrisoning the main routes, major cities, air bases, and logistics sites.
- Relieve the Afghan government garrison forces and push them back into the countryside to battle the resistance.
- Provide logistic, air, artillery, and intelligence support to Afghan forces.
- Provide minimum interface between Soviet forces and the local populace.
- Accept minimal Soviet casualties.
- Strengthen Afghan forces to defeat the resistance so that Soviet forces could withdraw.[72]

By December 22, the motor rifle divisions had arrived in their attack positions and paratroopers had secured the Kabul airport, the Salang Tunnel, and the road to Kabul. During December 24–26, a round-the-clock airlift moved 15,000 Soviet forces into Kabul and Bagram, including the 345th Guards Airborne Regiment with its BMDs. On the evening of the December 26, the Soviet forces began deploying to secure key locations.[73] Soviet advisers to the DRA Army were also busy. They "disabled equipment, blocked arms rooms and prevented a coordinated Afghan military response."[74] By Decem-

[70] Zaloga, *Inside the Blue Berets*, p. 230.

[71] Westermann, "The Limits of Soviet Airpower," p. 22.

[72] Grau and Yahya, "The Soviet Experience in Afghanistan," p. 19.

[73] Cordesman and Wagner, *The Afghan and Falklands Conflicts*, p. 33.

[74] Grau and Yahya, "The Soviet Experience in Afghanistan," p. 18.

ber 27, the invasion was in full swing. Soviet airborne and *Spetsnaz* forces seized key government installations and communications sites in Kabul, while "the Soviet ground invasion force crossed into the country, fought a few pockets of Afghan military resistance and occupied the main cities."[75]

On the evening of December 27, Soviet KGB *Spetsnaz* forces and paratroopers stormed the palace of President Amin. In the bloody firefight that followed, President Amin was killed. The Soviets then installed Babrak Karmal, whom they had transported to Kabul from his exile in Europe, as president. He obligingly "invited" the Soviet Army, already pouring into Afghanistan, to enter and help secure his country. In an attempt to win popular support, President Karmal released political prisoners from jail, included Khalq members in his regime, and announced "a moderate reform program and his respect for Islam."[76]

The Soviets secured the major cities and airfields in Afghanistan by early January 1980. It seemed, at least initially, that the earlier Soviet success in Czechoslovakia had been repeated in Afghanistan. This illusion was soon dispelled. The Soviets had "seriously miscalculated the willingness of both the Afghan Army and the Mujahideen to fight."[77] The DRA Army deserted in large numbers, often taking their weapons with them, while the Mujahideen began attacking Soviet units in the cities. The Mujahideen also began a protracted guerilla war against the Soviet occupiers and their DRA Army puppets. They were aided in their resistance by secure sources of materiel, supply, and sanctuary in Pakistan.[78]

The Soviet occupation of Afghanistan proceeded in four phases. The first phase lasted from the invasion until the initial consolidation of the occupation was complete. The second phase began in March

---

[75] Grau and Yahya, "The Soviet Experience in Afghanistan," p. 18.

[76] Cordesman and Wagner, *The Afghan and Falklands Conflicts*, p. 34.

[77] Westermann, "The Limits of Soviet Airpower," p. 18.

[78] Cordesman and Wagner, *The Afghan and Falklands Conflicts*, p. 96. See also Westermann, "The Limits of Soviet Airpower," pp. 62–64. China, Egypt, Iran, and, eventually, the United States funneled aid to the Mujahideen through Pakistan.

1980 and lasted until April 1985. During this period, the Soviets tried to crush the Mujahideen and secure the Pakistani border.[79] Operations became increasingly punitive, with Soviet and DRA forces conducting offensives that destroyed the homes and crops of villagers in areas that supported the Mujahideen.[80] The brief tenure of Konstantin Chernenko, General Secretary of the Communist Party of the Soviet Union from February 13, 1984, until his death on March 10, 1985, resulted in perhaps the most concentrated effort by the Soviets to win the war in this second phase. Secretary Chernenko led a shift "towards a policy of the 'iron fist' with respect to Afghanistan."[81]

Unfortunately for the Soviets, their counterinsurgency efforts were hampered by the need to divert substantial numbers of troops to protect critical LOCs, escort convoys, garrison outposts, and occupy major urban areas and 28 provincial capitals.[82] Eventually,

> [f]ully 80 percent of Soviet forces in Afghanistan conducted occupation duties, primarily involving security and support activities [for Soviet forces]. Counterinsurgency forces, consisting overwhelmingly of Slavic conscripts, constituted the remaining 20 percent of Russian ground forces and bore the brunt of combat operations.[83]

Over time, most of the counterinsurgency effort devolved to the more highly trained airborne, air-assault, reconnaissance, and *Spetsnaz* forces, although the regular motorized rifle units did "occasionally participate in conventional search and destroy missions and large sweeps against the resistance."[84]

---

79 Zaloga, *Inside the Blue Berets*, p. 250.

80 Cordesman and Wagner, *The Afghan and Falklands Conflicts*, p. 41.

81 Westermann, "The Limits of Soviet Airpower," p. 57.

82 Westermann, "The Limits of Soviet Airpower," p. 24.

83 Westermann, "The Limits of Soviet Airpower," p. 26.

84 Alexander Alexiev, *Inside the Soviet Army in Afghanistan*, Santa Monica, Calif.: RAND Corporation, R-3627-A, 1988, p. vi.

The Soviet force (which varied in number between 90,000 and 120,000 troops), was insufficient to secure its LOCs. It could not prevent the Mujahideen from interdicting the roads and pipelines critical to meeting the logistic needs of the Soviet and DRA forces.[85] Furthermore, the Soviets were attempting to control a large area that was roughly the size of Texas. The operational challenges presented by the sheer size of Afghanistan were further complicated by the nation's rugged terrain and poorly developed infrastructure, which in 1979 featured only 3,000 mi of all-weather roads and no rail lines.[86]

With the death of Secretary Chernenko from emphysema and the ascension to power of Mikhail Gorbachev in April 1985, Soviet policy entered its third phase—"Afghanization," or the shifting of responsibility for the antiguerilla campaign and large-scale ground fighting to the DRA Army.[87] By early 1986, Secretary Gorbachev was describing Afghanistan as a "bloody stump," and in May 1986, Najibullah Ahmadzai replaced Karmal as president. Secretary Gorbachev withdrew some Soviet forces in the summer of 1986.

The decision to extricate the Soviet Union from the Afghan quagmire was not reached until a November 13 Politburo meeting, however. Although fighting continued after that point, the Soviets were searching for a way out of the war that would leave the PDPA in control.[88] At at the end of 1987, the Soviets faced a situation in which their own

> and DRA forces had won most of their engagements in 1986 and 1987 but were gradually losing the war. In spite of a vast Soviet effort, the DRA (now the RA [Republic of Afghanistan]) still

[85] Grau and Yahya, "The Soviet Experience in Afghanistan," p. 23.

[86] Grau and Yahya, "The Soviet Experience in Afghanistan," pp. 19–20.

[87] Zaloga, *Inside the Blue Berets*, 250; Westermann, "The Limits of Soviet Airpower," p. 71; see also p. 87. "Afghanization" would prove difficult, because, "[d]espite the use of press gangs and extended enlistments, DRA regular forces never exceeded 40,000 during the entire period of the [Soviet] occupation."

[88] Westermann, "The Limits of Soviet Airpower," p. 80.

> lacked popular support at the beginning of 1988 and could not secure the countryside after any of its victories.[89]

One author concluded that the Soviet failure can be attributed to six factors: the availability of insurgent sanctuaries; the failure of Soviet interdiction efforts; the logistical parsimony of the Mujahideen; the small size of Soviet forces, especially counterinsurgency forces; the lack of appropriate counterinsurgency doctrine; and the Mujahideen's introduction of effective man-portable surface-to-air missile technology (which negated Soviet air supremacy).[90]

The fourth and final phase of the Soviet occupation began on February 8, 1988, when Secretary Gorbachev announced the Soviet withdrawal from Afghanistan. Soviet troops began their departure on May 15, 1988, and had completed the withdrawal by February 15, 1989. In the aftermath of the Soviet occupation, the Afghanis were left to resolve their civil war. Soviet losses for the nine-year war stood at

> 15,000 dead, 118 jet aircraft, 333 helicopters, 147 tanks, 1,134 armored personnel carriers, 1,138 communications and command post vehicles, 510 engineering vehicles and 11,369 trucks.[91]

The Afghan people also suffered. The Soviets had relied on massive firepower, largely delivered by "fixed-wing aircraft, helicopters, artillery,

---

[89] Cordesman and Wagner, *The Afghan and Falklands Conflicts*, p. 76.

[90] Westermann, "The Limits of Soviet Airpower," p. 104. See also George Crile, *Charlie Wilson's War: The Extraordinary Story of the Largest Covert Operation in History,* New York: Atlantic Monthly Press, 2003, for a discussion of the United States and other countries that armed and financed the Mujahideen resistance to the Soviet occupation of Afghanistan.

[91] Grau and Yahya, "The Soviet Experience in Afghanistan," p. 27; see also pp. 20–21:

> Far more telling were the 469,685 other [Soviet] casualties, fully 73 percent of the overall force, who were wounded or incapacitated by serious illnesses. Some 415,932 troops fell victim to disease—more than 115,300 contracted infectious hepatitis and 31,080 caught typhoid fever. These numbers point out that Soviet military hygiene and the conditions surrounding troop life were abominable. These diseases, unheard of in armies with modern medicine, had a staggering social impact on returnees and the Soviet population.

rocket launchers" to make up for their relatively low troop strength.[92] They used conventional munitions, chemical weapons, and mines to execute their strategy, resulting in an estimated 1.3 million Afghan civilian casualties. Additionally, the war produced "over five million refugees with over three million sheltered in Pakistan alone."[93]

The Armored Forces

The Soviet Union employed a wide variety of armored vehicles in Afghanistan, as shown in Table 3.4. Afghanistan's mountainous terrain and the enemy forces, however, made the use of tanks problematic and inappropriate in most cases, although they were available.[94] Soviet tanks, like the MBTs of almost all countries, were designed to fight other tanks in head-to-head fights. Consequently,

> Soviet tank guns did not prove particularly effective against most infantry targets in rough terrain. This was particularly true of the T-55, which lacked the gun elevation and depression to be effective in mountain combat. Soviet tanks lacked sights with the wide visual coverage and targeting aids necessary to easily locate typical Mujahideen targets or allow tank crews to deal rapidly with infantry ambushes. Commanders could not fight from open turrets, however, without taking losses. Tank guns could not be elevated to the degree required or tracked with sufficient speed, and crews experienced severe discomfort and fatigue problems because of the poor human engineering and ventilation of Soviet tanks.[95]

---

[92] Grau and Yahya, "The Soviet Experience in Afghanistan," p. 19. The Soviets used massive, often indiscriminate firepower. See Cordesman and Wagner, *The Afghan and Falklands Conflicts*, p. 134:

> [The Soviets] increasingly used bombers flying from the USSR [Union of Soviet Socialist Republics], and in November 1988, the Soviets started firing Scud missiles as a means of delivering both a terror weapon and fire support without exposing manned aircraft. Similarly, the USSR would use long-range artillery punitively to attack towns friendly to the Mujahideen, or captured by the Mujahideen, to force them to withdraw.

[93] Westermann, "The Limits of Soviet Airpower," p. 102.

[94] Grau and Yahya, "The Soviet Experience in Afghanistan," p. 24.

[95] Cordesman and Wagner, *The Afghan and Falklands Conflicts*, pp. 148–149. Cordesman and Wagner also note that, aside from issues of suitability as a weapon system vis-à-vis the

Table 3.4
Soviet Armored Vehicles in Afghanistan

Type	Weight (tons)	Armament	Max Armor (mm)
T-54/55 tank	39.00	100-mm gun; 12.7-mm machine gun; two 7.62-mm machine guns	203
T-62 tank	44.00	115-mm gun; 12.7-mm machine gun; 7.62-mm machine gun	242
T-64 tank	48.90	125-mm gun; 12.7-mm machine gun; 7.62-mm machine gun	N/A[a]
PT-76 reconnaissance vehicle	16.06	7.62-mm gun; 7.62-mm machine gun	14
D-1 airborne combat vehicle	8.50	73-mm gun; three 7.62-mm machine guns	23
BMD-2	11.50	30-mm cannon; two 7.62-mm machine guns; ATGM	10
BMP-1 IFV	14.80	73-mm gun; 7.62-mm machine gun; ATGM	33
BMP-2 IFV	15.70	30-mm cannon; 7.62-mm machine gun; ATGM	N/A[b]
BTR-D	6.70	7.62-mm machine gun; many variants[c]	N/A[d]
BTR-60PB APC with armored roof	11.40	14.5-mm machine gun or 12.7-mm machine gun, and up to three 7.62-mm machine guns	9
BTR-70	12.65	14.5-mm machine gun; 7.62-mm machine gun variants with AGS-17 30-mm automatic grenade launcher (AGL) and/or machine guns	9
BTR-80	14.96	14.5-mm machine gun; 7.62-mm machine gun	9

SOURCES: Foss, *Jane's Tanks and Combat Vehicles Recognition Guide*, pp. 66–75, 172–181, 410–417; Federation of American Scientists, "Military Analysis Network—U.S. Land Warfare Systems," n.d.; Cordesman and Wagner, *The Afghan and Falklands Conflicts*, pp. 148–154. Cordesman and Wagner note on p. 154 that the Mujahideen, although they had captured armored systems, used armor

> defensively and largely in the rear. Captured tanks were used largely as artillery weapons. This was not a critical problem [for the Soviets] during most of the war, since main force concentrations could not survive. The

Table 3.4—Continued

> Mujahideen found, however, that they could not assault many strong points after the beginning of the Soviet withdrawal in 1988 because they lacked armored mobility and protection in the assault phase.

See also pp. 12–13, where Cordesman and Wagner note that the DRA Army was equipped with the following Soviet equipment: T-34, T-54/55, and T-62 tanks; PT-76 light tanks; BMP-1 armored fighting vehicles; and BTR-40 and BTR-152 APCs.

[a] Classified (laminate, steel, or reactive).

[b] Classified.

[c] See Zaloga, *Inside the Blue Berets*, pp. 173–176. The Soviet army employed several specialized variants of the BTR in its airborne units: the D2S9 120-mm self-propelled combination gun, with a turreted breech-loading mortar/howitzer system; the BMD-KShM C2 variant; the BREM-D armored repair and recovery variant; the BTR-RD ATGM variant; and the BTR-ZD air defense variant with a towed ZU-23 twin 23-mm air defense gun.

[d] Antibullet (7.62 mm).

During the phase of the war when the Soviets assumed the major burden of the fighting, they relied on lighter armored fighting vehicles (AFVs) and APCs. This gave Soviet forces "a higher degree of maneuverability and effectiveness where problems with ambushes were especially acute."[96] Later in the war, tanks were mainly relegated to the role of support guns and "avoided direct combat."[97] Thus, medium-armored vehicles provided much of the Soviet armor capability throughout the war in Afghanistan.

Employment

The Soviet forces that deployed to Afghanistan were organized, trained, and equipped for conventional warfare against similar conventional forces (those of NATO [North Atlantic Treaty Organization], for example).[98] At the beginning of the war, the Soviets attempted to apply their conventional doctrine to the conflict in Afghanistan. They used conventional combined-arms formations, supported by air and artil-

---

[96] Cordesman and Wagner, *The Afghan and Falklands Conflicts*, p. 149.

[97] Cordesman and Wagner, *The Afghan and Falklands Conflicts*, p. 149.

[98] Grau and Yahya, "The Soviet Experience in Afghanistan," p. 21.

lery, in large-scale, set-piece operations to find and destroy the Mujahideen. These tactics were ineffective because the Soviet operations

> were slow moving, armor heavy, road bound, and poorly executed. The resulting "convoy mentality" made Soviet armor vulnerable to Mujahideen anti-tank rocket launchers and led to numerous successful ambushes of Soviet and Afghan forces.[99]

In mountainous terrain, Soviet armored units experienced similar problems. They were, given the restrictive terrain, road-bound. Additionally, Soviet soldiers tended to try to fight from their vehicles, shooting through firing ports rather than dismounting. Furthermore, they did not secure the high ground along convoy routes or conduct adequate ground reconnaissance. In short, they were vulnerable to Mujahideen ambush.[100]

As the war progressed, the Soviets adapted their tactics and equipment. Instead of large-scale operations with conventional units, they increasingly relied on airmobile, motorized rifle, airborne, and *Spetsnaz* units. They also "stopped set piece attacks and started hit-and-run operations. They emphasized the use of lighter APCs and AFVs, including the use of BMDs lifted by Mi-6 helicopters."[101]

The Soviets had learned that lighter armored vehicles were more useful in counterinsurgency, particularly in mountainous terrain. They began to rely more heavily on BMDs for service in the mountains because tanks and BMPs "were too large for mountain trails, subject to frequent breakdowns, and difficult to service in the field."[102] Larger

---

[99] Cordesman and Wagner, *The Afghan and Falklands Conflicts*, p. 125.

[100] Cordesman and Wagner, *The Afghan and Falklands Conflicts*, p. 124.

[101] Cordesman and Wagner, *The Afghan and Falklands Conflicts*, p. 124.

[102] Cordesman and Wagner, *The Afghan and Falklands Conflicts*, p. 150. See also Zaloga, *Inside the Blue Berets*, p. 241. Zaloga notes the following:

> The BMD-1 airborne assault vehicle proved a disappointment in combat. . . . Its suspension had been designed for light weight and as a result was very fragile. It soon became chewed up in Afghanistan's rocky terrain. It was too cramped for sustained operations, and its 73mm Grom low-pressure gun could not elevate enough to reach the mujahideen high in the mountains. Its small size made it very vulnerable to mine damage, and the

vehicles, including tanks, BMPs, BTR-60s, BTR-70s, and BTR-80s, were used "in the cities, for LOC protection, as armored ambulances, and in defending strategic crossroads." The Soviets also modified BMDs, BMPs, and other vehicles to make them more effective against infantry targets than against the armored vehicles of NATO that they had been designed to fight. Rapid-fire, hyperelevating 30-mm cannons replaced the 76-mm guns on BMDs and BMPs. Other vehicles mounted hyperelevating machine guns or AGS-17 rapid-fire grenade launchers. These new weapons provided

> machine guns and cannon which could "hose" targets with high rates of fire, easily track targets (unlike slow-moving heavy guns), and be easily hyper-elevated . . . to provide rapid surge fire and direct fire support.[103]

Finally, vehicles often deployed smoke canisters to screen themselves when ambushed.[104]

Beginning in 1984, the Soviets improved the training of their infantry and gave it greater dismounted firepower. AFVs were also used more effectively. The Soviets created the *bronegruppa* concept that separated the infantrymen from their combat vehicles. The vehicles were then used in a separate, supporting role

> that could attack independently on the flanks, block expected enemy withdrawal routes, serve as a mobile fire platform to reinforce elements in contact, perform patrols, serve in an economy-of-force role in both offense and defense and provide convoy escort and security functions.[105]

---

paratroopers soon learned the Vietnam lesson that it was safer to ride outside of a vehicle than inside when mines were present.

[103] Cordesman and Wagner, *The Afghan and Falklands Conflicts*, p. 151.

[104] Cordesman and Wagner, *The Afghan and Falklands Conflicts*, p. 153.

[105] Grau and Yahya, "The Soviet Experience in Afghanistan," p. 23.

Key Insights

The Soviet experience in Afghanistan offers several overarching insights about using medium-armored forces to rapidly topple a regime and in a protracted counterinsurgency:

- Rapidly deployable medium-armored forces were a critical capability in the Soviet coup de main operation in Afghanistan.
- In the type of warfare the Soviets experienced in Afghanistan, medium-armored vehicles provided essential protected mobility and firepower in difficult terrain during combat operations and for convoy support. Soviet tanks proved unsuitable for these roles, and light infantry was vulnerable. The relatively light armor of Soviet medium-armored platforms made the platforms vulnerable to mines, RPGs, and heavy machine guns.
- The combat vehicle weapons needed in the conflict environment the Soviets faced in Afghanistan—a fight against light guerilla forces in mountainous terrain—differed from those weapons that most Western nations put on their combat vehicles, which were usually designed for direct-fire engagements against other armored vehicles. In Afghan-like conflicts, armored-vehicle weapons need to be rapid-firing, hyperelevating, and capable of suppressive fire. This is true for any platform operating in this type of environment. The Soviets eventually settled on medium-armored platforms because of their greater mobility and sustainability over difficult terrain and protracted distances.

U.S. Forces in Operation Just Cause (Panama, 1989)

U.S. concerns with the regime of Panamanian dictator Manuel Antonio Noriega began to manifest themselves in 1985. In that year, U.S. National Security Adviser John M. Poindexter and Assistant Secretary of State for Inter-American Affairs Elliot Abrams warned Noriega "of U.S. concern over his monopoly of power and involvement in the drug trade."[106] In June 1987 Noriega brutally suppressed unarmed demonstrations against his regime, and in September 1987 the U.S. Senate passed a resolution calling on Noriega to step down from power. In November 1987 the United States stopped economic and military aid to Panama in the aftermath of an attack on the U.S. embassy in Panama.[107]

Tensions heightened on February 5, 1988, when U.S. federal grand juries handed down drug trafficking indictments on Noriega and several of his key lieutenants. That same month, the United States initiated contingency planning (OPLAN Blue Spoon) for an intervention in Panama.[108] For his part, Noriega began a campaign of harassment against U.S. citizens in Panama and turned to Cuba, Nicaragua, and Libya for aid.[109] In response, the United States deployed additional forces to provide security to U.S. installations and began moving military dependents and diplomats out of Panama.[110]

Tensions in Panama increased in May 1989 when Noriega voided the results of an election in which the candidates of his regime lost the popular vote. In the aftermath of the election, opposition leaders were assaulted and many went into hiding. U.S. President George H. W. Bush responded by ordering the deployment of some 1,800 U.S. troops to Panama. This deployment, termed Operation Nimrod Dancer, included a brigade headquarters and light infantry battalion task force from the 7th Infantry Division, a mechanized infantry battalion from the 5th Infantry Division, and a U.S. Marine Corps light-armored

---

[106] Ronald H. Cole, *Operation Just Cause: The Planning and Execution of Joint Operations in Panama, February 1988–January 1990*, Washington, D.C.: Joint History Office, 1995, p. 6.

[107] Cole, *Operation Just Cause*, p. 6.

[108] Center for Army Lessons Learned [CALL], *Operation Just Cause Lessons Learned*: Vol. I, *Soldiers and Leadership* (No. 90-9), Fort Leavenworth, Kan.: U.S. Army Combined Arms Command, 1990, pp. ii, 199.

[109] Cole, *Operation Just Cause*, p. 6.

[110] CALL, *Operation Just Cause Lessons Learned*, Vol. I, pp. ii, I–4.

infantry company. These forces provided additional security to U.S. citizens in Panama and conducted freedom of movement exercises.[111]

The Bush administration began to move toward a tougher Panama policy in the spring and summer of 1989, culminating in the promulgation of National Security Directive 17. The directive authorized actions "to assert U.S. treaty rights in Panama and to keep Noriega and his supporters off guard."[112] These measures included the publicized evacuation of U.S. dependents, increased patrolling, military exercises, and increased U.S. troop reconnaissance and armed convoys near Panamanian Defense Force (PDF) installations. Additionally, plans were made to take over key facilities from the PDF.[113]

In the aftermath of an October 1, 1989, failed coup attempt by PDF Major Moisés Giroldi, the pace of U.S. contingency planning accelerated. In early November, Lieutenant General Carl Stiner's XVIII Airborne Corps, the headquarters designated to lead Joint Task Force South (JTFSO), completed OPLAN 90-2 for the invasion of Panama. At the heart of the plan was the rapid neutralization of the PDF through the simultaneous strike of 27 targets.[114] U.S. forces in Panama also increased their activity. As Secretary Cheney recalled,

> [w]e adopted a more aggressive posture. We sent U.S. forces up and down the causeways, conducted helicopter operations, and scheduled exercises . . . . [These exercises] increased the tension and incidents between U.S. and PDF troops; and, by their frequency,

---

[111] CALL, *Operation Just Cause Lessons Learned*, Vol. I, pp. ii, I–4; Cole, *Operation Just Cause*, pp. 10–11. See also Nicholas E. Reynolds, *Just Cause: Marine Operations in Panama 1998–1990*, Washington, D.C.: Headquarters, U.S. Marine Corps, 1996, p. 15. The Carter-Torrijos Treaty, which turned the Panama Canal over to Panama, guaranteed the freedom of movement of U.S. forces in and around the canal to provide for its defense.

[112] Cole, *Operation Just Cause*, p. 12.

[113] Cole, *Operation Just Cause*, p. 12.

[114] CALL, *Operation Just Cause Lessons Learned*, Vol. I, p. I–4; see also p. I–5. OPLAN 90-2 had the following objectives: protect U.S. lives and key sites and facilities, capture and deliver Noriega to competent authority, neutralize PDF forces, neutralize PDF C2, support establishment of a U.S.-recognized government in Panama, and restructure the PDF.

> caused Noriega to believe that the U.S. was trying to intimidate him. Consequently, Noriega did not expect an attack.[115]

On December 15, 1989, the Panamanian National Assembly passed a resolution declaring "that 'owing to U.S. aggression,' a state of war existed with the United States," and Noriega named himself "Maximum Leader."[116] Over the next two days, incidents against U.S. military personnel and dependents increased. On September 16, PDF forces fired on four U.S. officers in their car, wounding three of them, one of whom eventually died. PDF forces also arrested a U.S. naval officer and his wife, and abused both while detaining them. After intensive consultations between U.S. civilian and military leaders, President Bush authorized execution of Operation Just Cause on December 18, with D-Day and H-Hour set at December 20, 1:00 a.m. Panama time. Based on increased PDF activity, however, General Stiner advanced H-Hour by fifteen minutes.[117]

Operation Just Cause was a joint operation executed by a very straightforward command structure. General Maxwell Thurman, Commander in Chief, Southern Command, was the overall commander. General Stiner, the commanding general of XVIII Corps, had operational control of the fighting forces as the commander of JTFSO. General Stiner commanded four conventional task forces (Semper Fi, Atlantic, Pacific, and Bayonet), the air component command, and the joint special operations task force (JSOTF) commanded by Major General Wayne A. Downing. General Downing's JSOTF controlled six special operations task forces (Red, Green, Black, Gator, Blue, and White).[118]

---

[115] Cole, *Operation Just Cause*, p. 24.

[116] Cole, *Operation Just Cause*, p. 27.

[117] Cole, *Operation Just Cause*, pp. 27–33; CALL, *Operation Just Cause Lessons Learned*, Vol. I, p. I–4. See also Cole, *Operation Just Cause*, pp. 37–38. At H-Hour, the U.S had some 13,000 troops in Panama. During Just Cause, this number rose to approximately 27,000 troops, of which 22,000 were engaged in combat operations.

[118] Jennifer Morrison Taw, *Operation Just Cause: Lessons for Operations Other Than War*, Santa Monica, Calif.: RAND Corporation, MR-569-A, 1996, pp. 9–10, 35–36.

U.S. forces rapidly accomplished their operational objectives—with the notable exception of capturing Noriega, who had taken refuge in the papal nuncio in Panama City—and combat operations ended on December 26.[119] On January 4, Noriega surrendered to U.S. forces and was turned over to Drug Enforcement Administration agents, who arrested him. At 9:31 p.m., Noriega departed Panama for Florida aboard a U.S. Air Force C-130. All the objectives of Just Cause had been accomplished. By January 4, the redeployment of U.S. forces had begun. The Joint Staff terminated Operation Just Cause on January 11, 1990.[120]

The Armored Forces

Given the paucity of armor in the PDF, the U.S. deployed relatively few armored vehicles during Operation Just Cause. Nevertheless, these vehicles made a direct contribution to the success of the operation. The armored vehicles deployed during Operation Just Cause included M113 APCs, M551 Sheridan armored reconnaissance airborne assault vehicles, and three LAV variants. These medium-armored platforms were allocated to Task Force Bayonet, Task Force Pacific, and Task Force Semper Fi. Table 3.5 describes the characteristics of the U.S. armored vehicles used in Operation Just Cause. The PDF's armored vehicles are shown in Table 3.6.

Task Force Bayonet, commanded by the 193rd Infantry Brigade, included the 4th Battalion, 6th Infantry, from the 5th Infantry Division (Mechanized), who were equipped with M113 APCs. The mission of Task Force Bayonet included isolating and clearing the *Commandancia* (PDF headquarters), seizing and securing key installations and defense sites in and around Panama City, and protecting U.S. housing areas. The units of the 4th Battalion, 6th Infantry, were in two subordinate task forces, Task Force Gator and Task Force Wildcat. Task Force Gator was commanded by the 4th Battalion, 6th

[119] Cole, *Operation Just Cause*, p. 65.

[120] Cole, *Operation Just Cause*, pp. 57–63, 69.

Table 3.5
U.S. Armored Vehicles in Panama

System	Weight (tons)	Armament	Max Armor (mm)
M113 APC	12.5	.50-caliber machine gun	38
M551 Sheridan armored reconnaissance airborne assault vehicle[a]	17.4	152-mm gun/missile launcher; .50-caliber machine gun; 7.62-mm machine gun	N/A[b]
LAV-25	14.1	25-mm chain gun; 7.62-mm machine gun	10
LAV-L	14.1	7.62-mm machine gun	10
LAV-C2	13.53	7.62-mm machine gun	10

SOURCES: LAV data from Foss, *Jane's Tanks and Combat Vehicles Recognition Guide*, and U.S Marine Corps, Headquarters, homepage.

[a] See also Frank Sherman, "Operation Just Cause: The Armor-Infantry Team In the Close Fight," *Armor*, Vol. 105, No. 5, September–October 1996, p. 34. Sherman notes that the M551 Sheridans used in Operation Just Cause had a modified version of the M60A3 tank thermal sight, giving it an excellent night fighting capability—a significant improvement over versions of the Sheridan used in Vietnam.

[b] Aluminum hull; steel turret.

Table 3.6
PDF Armored Vehicles in Panama

System	Weight (tons)	Armament	Max Armor
V-150	10.5	Several versions with machine guns	Classified[a]
V-300	16.16	90-mm gun; 7.62-mm machine gun	Classified[a]

SOURCES: Foss, *Jane's Tanks and Combat Vehicles Recognition Guide*, p. 392; Foss, *Jane's World Armoured Fighting Vehicles*, pp. 290–292; Federation of American Scientists, "Military Analysis Network—U.S. Land Warfare Systems." See also International Institute for Strategic Studies, *The Military Balance: 1989–1990*, London: Brassey's, 1989, p. 198, which notes that the PDF had sixteen V-150 and thirteen V-300 vehicles.

[a] Capable of stopping a 7.62-mm ball.

Infantry (-),[121] and included a platoon from the 3rd Battalion, 73rd Armor, 82nd Airborne Division, who were equipped with M551 Sheridans, and a platoon of Marines from Company D, 2nd Light Armored Infantry Battalion, who were equipped with LAVs. Task Force Gator had the mission of taking the *Commandancia*. Task Force Wildcat was commanded by the 5th Battalion, 87th Infantry, and included the M113-equipped Company A, 4th Battalion, 6th Infantry. Its mission was the neutralization of several installations and the emplacement of roadblocks to isolate the *Commandancia*.

Task Force Pacific, commanded by the 82nd Airborne Division, included Company C (-), 3rd Battalion, 73rd Armor, 82nd Airborne Division. Task Force Pacific's missions included conducting a parachute assault on the Torrijos airport, isolating and neutralizing PDF forces at Panama Viejo, Tinajitas, and Fort Cimarron, and denying reinforcement of Panama City by PDF units in its area of operations.[122]

Task Force Semper Fi, commanded by Marine Forces Panama, included Company D, 2nd Light Armored Infantry Battalion (-). Task Force Semper Fi was responsible for blocking the western approaches into Panama City and securing the Bridge of the Americas. On December 20, the task force received an additional assignment to attack and secure the headquarters of the 10th Military Zone in La Chorrera. The Marine Security Guard Detachment attached to the U.S. Embassy in Panama City remained responsible for embassy security.[123]

The United States employed medium-armored vehicles during Operation Just Cause. They were medium-armored when compared to the light infantry and heavy-armored vehicles in the U.S. Army and U.S. Marine Corps.[124]

---

[121] The *(-)* symbol indicates that the company, battalion, or division deployed at less than full strength.

[122] Taw, *Operation Just Cause: Lessons for Operations Other Than War*, pp. 7–8, 35–36; CALL, *Operation Just Cause Lessons Learned*, Vol. I, pp. I-9–12.

[123] Taw, *Operation Just Cause: Lessons for Operations Other Than War*, pp. 7, 35–36; Reynolds, *Just Cause: Marine Operations in Panama*, pp. 22–26.

[124] See Thomas Donnelly, Margaret Roth, and Caleb Baker, *Operation Just Cause: The Storming of Panama*, New York: Lexington Books, 1991, pp. 74, 230, 345. The PDF's V-150 and

Employment

Task Force Bayonet. The medium-armored forces in Task Force Bayonet had deployed to Panama prior to Operation Just Cause. The 4th Battalion, 6th Infantry, and Company D, 2nd Light Armored Infantry Battalion, were part of the Operation Nimrod Dancer deployments in May 1989.[125] The M551 Sheridan platoon entered Panama surreptitiously, flying into Howard Air Force Base and moving into hidden positions before Operation Just Cause.[126]

Task Force Gator's medium-armored forces made a significant contribution to the assault on the *Commandancia,* which began at 12:45 a.m. on December 20. M551 Sheridans and LAV 25s—as well as AC-130 gunships and AH-64 Apache helicopters—supported the assaulting B Company, 4th Battalion, 6th Infantry, with fires. After an intense firefight, B Company broke through roadblocks around the *Commandancia* and moved into blocking positions. Infantry assaults finally enabled U.S. forces to take the *Commandancia* by 6:00 p.m. that evening.[127] M551s were particularly useful as protected mobile-gun platforms and "provided direct fire support from overwatch positions and blew entry holes in buildings for infantry assaults."[128]

Task Force Pacific. Ten M551 Sheridans made the airborne assault on the Torrijos airport as part of Task Force Pacific. Of the ten vehicles, one was destroyed in the jump and another was so badly damaged that it "became a source of parts to keep the others going."[129]

V-300 Cadillac Gage armored cars had little impact on the operation. In general, those vehicles that encountered U.S. forces fled or were destroyed by AT-4 antitank weapons or fire (e.g., Hellfire missiles) from attack helicopters.

[125] Cole, *Operation Just Cause,* p. 11; Reynolds, *Just Cause: Marine Operations in Panama,* pp. 14–15. Company D, 2nd Light Armored Infantry Battalion, arrived with 14 LAV-25s, two LAV-Ls, one LAV-C2, and ten four-man scout teams.

[126] Sherman, "Operation Just Cause: The Armor-Infantry Team in the Close Fight," p. 34.

[127] Cole, *Operation Just Cause,* p. 41.

[128] Center for Army Lessons Learned (CALL), *Operation Just Cause Lessons Learned*: Vol. III, *Intelligence, Logistics & Equipment* (No. 90-9), Fort Leavenworth, Kan.: U.S. Army Combined Arms Command, 1990, pp. III-14–15.

[129] Daniel P. Bolger, *Death Ground: Today's American Infantry in Battle,* Novato, Calif.: Presidio Press, 2000, p. 47.

The eight remaining usable Sheridans provided valuable support to the 82nd Airborne's light infantry, serving as mobile guns to reduce enemy positions and to clear roadblocks. Additionally, Sheridans were able to cross Panamanian bridges that "could not support the Army's heavier main battle tank."[130] Finally, they provided overwatch fires for infantry operating in areas whose densely forested character made the use of AC-130 gunships impractical.[131]

Task Force Semper Fi. The 13 LAVs of Company D, Light Armored Infantry Battalion, began operations on D-Day by taking the PDF's Directorate of Traffic and Transportation Station No. 2 in Arraijan. They also provided supporting fires to U.S. Marine Corps infantrymen. Early on D-Day, Task Force Semper Fi had accomplished its missions of securing the Bridge of the Americas and blocking the western approaches to Panama City. Company D was also used to assault and secure the headquarters of the 10th Military Zone in La Chorrera. The protected firepower and mobility of the LAVs enabled the task force to reduce roadblocks en route to the objective and take the PDF compound.[132]

Key Insights

Operation Just Cause is a case where jungles, limited infrastructure, dispersed operations, and urban operations made light infantry the most appropriate force and made the use of heavy armor impracticable. Additionally, Operation Just Cause points to the need for air-droppable armor to support airborne and ranger forced-entry operations. Other insights include the following:

- Although Operation Just Cause did not rigorously test the potential of medium-armored forces, these forces did contribute to the rapid success of the operation. They provided protected mobility

---

[130] R. Cody Phillips, *Operation Just Cause: The Incursion into Panama*, Washington, D.C.: U.S. Army Center of Military History, n.d., p. 45.

[131] Sherman, "Operation Just Cause: The Armor-Infantry Team in the Close Fight," pp. 34–35.

[132] Reynolds, *Just Cause: Marine Operations*, pp. 22–26.

and fire support that was an important complement to U.S. light forces.

- In an operational environment characterized by highly restrictive rules of engagement and objectives suited to light infantry, the armored platforms provided fire support, protected mobility, and blocking forces.
- The air-droppable M551 Sheridan's role as a mobile assault gun is particularly noteworthy, providing as it did critical capabilities to forced-entry airborne and ranger forces. The Sheridans reduced enemy roadblocks and breached walls in built-up areas. Both of these uses enabled light infantry to maneuver in places where it had been stymied before the employment of the Sheridan. Heavy-armored vehicles (i.e., tanks) would likely not have been available to fulfill all of these roles: Many Sheridans had to be air-dropped with paratroopers, and heavy-armored vehicles could not have been transported in this manner.
- Medium armor was able to traverse Panamanian bridges that would not have supported heavy armor.

Russia in Chechnya (1994–2001)

Russia and Chechnya have a long—and often troubled—history. Russia seized control of the majority of Chechen territory during the Czarist era in the 19th century (after some 30 years of fighting) but was "never able to fully incorporate the Chechen people into the Russian empire."[133] In the aftermath of the Bolshevik Revolution, the Soviets also had their share of trouble with the Chechens. During World War II, a portion of the Chechens sided with the Nazis, hoping to gain independence from the Soviets. Soviet Union Secretary Joseph Stalin's retribution for this "betrayal" was typically brutal. He shipped the entire Chechen population to Kazakhstan, killing some 25 percent in the process. In the late 1950s, Secretary Nikita Khrushchev allowed

[133] Raymond C. Finch, III, "A Face of Future Battle: Chechen Fighter Shamil Basayev," *Military Review*, Vol. 77, No. 3, May–June 1997, pp. 33–41.

the Chechens to return to their homeland, where they lived largely uneventfully until the collapse of the Soviet Union. Then, once again, the Chechens sought independence.[134]

Chechnya I

The events leading to the first Russian intervention in Chechnya in 1994 began in the wake of the turmoil caused by the dissolution of the Soviet Union. In 1991, Dzhokhar Dudayev, a former Soviet Air Force general and ethnic Chechen, rose to lead the Chechen popular congress. In October 1991 he declared Chechnya's independence.[135] Although it opposed Chechen independence, Russia was too busy dealing with internal issues to do much about Chechnya.[136]

Russia did, however, support an opposition movement by the Ingush, the second-largest nationality in Chechnya. In November 1994,

> [a] force of 5,000 Chechen [Ingush] rebels and 85 Russian soldiers with 170 Russian tanks attempted to overthrow the Chechen government with a *coup de main* by capturing Grozny 'from the march' as they had in years past captured Prague and Kabul. They failed and lost 67 tanks in city fighting.[137]

In the aftermath of the abortive coup, the Ingush asked Russian Federation President Boris Nikolayevich Yeltsin for a full-scale Russian intervention. President Yeltsin responded on November 29, 1994, with decree Number 2137c, "On Steps to Reestablish Constitutional Law

[134] Finch, "A Face of Future Battle."

[135] Chad A. Rupe, "The Battle of Grozny: Lessons for Military Operations on Urbanized Terrain," *Armor*, Vol. 108, No. 3, May–June 1999, p. 20.

[136] Finch, "A Face of Future Battle."

[137] Lester M. Grau, "Russian Urban Tactics: Lessons from the Battle for Grozny," *National Defense University Strategic Forum 38*, 1994. See also Timothy Thomas, "The Battle of Grozny: Deadly Classroom for Urban Combat," *Parameters*, Vol. 29, No. 2, Summer 1999, pp. 87–102. Thomas notes that the Russians initially denied their involvement in the attempted Ingush coup, but recanted when the Chechens "paraded several captured Russian soldiers before TV cameras."

and Order In the Territory of the Chechen Republic." The rationale for the decree, as stated in the document, included

> the blatant violations of the Constitution of the Russian Federation in the Chechen Republic, the refusal of D. Dudaev to seek a peaceful resolution to the crisis, the increase in general criminal activity, continuing violations of the rights of and freedoms of citizens, repeated incidents of hostage-taking, and the increasing numbers of murders.[138]

The formation of a special group to plan and direct operations in Chechnya, headed by Minister of Defense Pavel Grachev, followed President Yeltsin's degree. Minister Grachev had the authority to create a "Joint Grouping of Federal Forces" designed to achieve the following goals:

- Stabilize the situation in the Chechen Republic.
- Disarm illegal armed bands and, in the event of resistance, destroy them.
- Reestablish law and order in the Chechen Republic in accordance with legislation of the Russian Federation.[139]

The planning group developed a four-stage operation to accomplish these goals. Stage one (November 29–December 6) involved creating and assembling "force groupings for operations towards Mozdok, Vladikavkaz, and Kizliar." In stage two (December 7–9), Russian forces planned to

---

[138] Anatoliy S. Kulikov, "The First Battle of Grozny," in Russell W. Glenn, ed., *Capital Preservation: Preparing for Urban Operations in the Twenty-First Century: Proceedings of the RAND Arroyo-TRADOC-MCWL-OSD Urban Operations Conference, March 22–23, 2000*, Santa Monica Calif.: RAND Corporation, CF-162-A, 2001, p. 17. Dzhokhar Dudayev's name is spelled several different ways in various accounts on Chechnya.

[139] Kulikov, "The First Battle of Grozny," in Glenn, ed., *Capital Preservation*, pp. 18–19, 45. General Anatoly Sergeevich Kulikov was a member of the special planning group headed by Minister Grachev, and as Commander in Chief, Internal Troops, of Russia's Interior Ministry, was the commander of Russian forces in Chechnya following their capture of Grozny.

> advance on Grozny from six directions and blockade it by forming two concentric rings. The outer ring was to coincide with the administrative border of Chechnya and the inner ring with the outside limits of the city of Grozny.

During stage three (December 10–13), Russian Army units, advancing from the north and south, would capture Grozny and its critical nodes, namely the presidential palace and key government buildings. They were then to begin disarming illegal formations. Finally, stage four (December 15–23) envisioned that Russian armed forces would continue to stabilize the situation while a transfer of responsibility to Internal Affairs Ministry forces, who would confiscate weapons and disarm rebel bands, occurred. On November 29, the National Security Council of the Russian Federation approved the four-stage plan. General-Colonel A. N. Mityukhin was given command of the Joint Grouping of Forces on December 5.[140]

The Russians were optimistic about their prospects for a quick victory in Grozny. Indeed, Minister Grachev

> had boasted . . . that he could seize Grozny in two hours with one parachute regiment. So the Russians drove into Grozny expecting to capture the city center and seat of government with only token resistance.[141]

---

[140] Kulikov, "The First Battle of Grozny," in Glenn, ed., *Capital Preservation*, pp. 22–37, quote on p. 24. There is an obvious gap in the stages—December 14 is missing.

[141] Grau, "Russian Urban Tactics." See also Thomas, "The Battle for Grozny," p. 2. Thomas notes that Grachev may have been putting on a good public face and that he actually harbored deep concerns about the Russian prospects in Grozny:

> The Russian armed forces that attacked Grozny, while well-equipped, were not the same professional force that opposed the West during the Cold War. Russian Minister of Defense Pavel Grachev, in a top-secret directive, listed some of the problems of his armed forces just ten days before the start of the war. He noted that the combat capabilities of the armed forces were low, the level of mobilization readiness was poor, and the operational planning capability was inadequate. Soldiers were poorly trained. Their suicide rates as well as the overall number of crimes in the force were up. Knowing the situation so clearly, Grachev's bold prediction that he could take Grozny with a single airborne regiment in two hours is incomprehensible. Perhaps Grachev privately understood the

The operational plan, however, proved unrealistic, and the Chechen opposition was much more capable than expected.[142]

To begin with, the hastily assembled Russian forces could not meet the timetable. Only one of the six groups reached its position according to plan; the other five were not in place until December 21. Consequently, the blockade of Grozny was not realized and the south of the city remained open to the rebel movement. Nevertheless, the Russian National Security Council decided on December 26 to take Grozny.[143]

On December 31, the Russians attacked the city. Although they had assembled a force of some 23,800 troops for operations in Chechnya, only 6,000 were used in the initial assault on Grozny.[144] The plan was for "storm detachments" to move in three groupings and attack the city simultaneously from the north, west, and east. The Russians had high expectations for the operation:

> [F]ederal forces, approaching a single point from three directions, would fully surround Dudaev's forces located in the center of the city. Casualties among the Russian troops would be minimized, as would collateral damage to the city of Grozny.[145]

---

true problems in the force but put on the face of public bravado to support the presidential directive he had received.

[142] Lester W. Grau, "Technology and the Second Chechen Campaign: Not All New and Not That Much," in Aldis, Anne, ed., *Strategic and Combat Studies Institute Occasional Paper No. 40: The Second Chechen War*, Shrivenham, UK: Conflict Studies Research Centre, 2000, p. 105.

[143] Grau, "Technology and the Second Chechen Campaign" in Aldis, ed., *Strategic and Combat Studies Institute Occasional Paper No. 40*, pp. 37–38.

[144] See Rupe, "The Battle of Grozny," p. 20, where Rupe notes the following:

> In December 1994, the Russian Army assembled three army groups consisting of 23,800 soldiers and special police units equipped with 80 tanks (T-72s, T-80s), 208 IFVs and APCs (BMP-2s, BMDs, BTR-70s), and 180 guns and mortars.

Rupe also cautions that these numbers "vary depending on the report."

[145] Rupe, "The Battle of Grozny," pp. 39–43, quote on p. 43. See also Grau, "Russian Urban Tactics." Grau describes a "storm detachment" as

Furthermore, there was a certain arrogance in the Russian plans: "When the Russian columns advanced into the center of Grozny, the men expected to disband poorly trained civilian mobs through a show of force by the Russian Army."[146]

The rebels soon disabused the attacking Russians of their expectations of a quick and easy victory. About 15,000 Chechen fighters awaited the Russians in Grozny in prepared defenses.[147] These defenses were organized in three lines:

> [The] outer and middle defense lines were based on strongpoints while the inner line consisted of prepared positions for direct artillery and tank fire. Lower and upper floors of buildings were prepared for fire from firearms and antitank weapons.[148]

> usually a motorized rifle battalion reinforced with at least a battalion of artillery, a tank company, an engineer company, an air defense platoon, flamethrower squads and smoke generator personnel. Artillery and air support are available from division assets.

In the case of the battle for Grozny, "their formation was often counterproductive because it destroyed what unit integrity existed in platoons, companies and battalions and gave commanders more assets than they could readily deploy and control."

[146] Rupe, "The Battle of Grozny," p. 21. Russian soldiers had been told to expect this reaction by their commanders, who in turn believed it based on earlier experiences in Prague and Kabul.

[147] See Rupe, "The Battle of Grozny," p. 20. Rupe notes that the Chechens had "15,000 personnel . . . with 60 guns and mortars, 30 Grad multiple rocket launchers, 50 tanks (most were non-operational), 100 IFVs, and 150 anti-aircraft guns."

[148] Timothy Thomas, "The Caucasus Conflict and Russian Security: The Russian Armed Forces Confront Chechnya: Military Activities of the Conflict During 11–31 December 1994," *The Journal of Slavic Military Studies*, Vol. 8, No. 2, June 1995. Here, Thomas describes the defensive lines:

> [T]he Chechen command created three defense lines to defend Grozny: an inner one with a radius of 1–1.5 km around the Presidential Palace; a middle one to a distance of up to 1 km from the inner borderline in the northwestern part of the city and up to 5 km in its southwestern and southeastern parts; and an outer border that passed mainly through the city outskirts.

Additionally, the Chechens intended to employ hit-and-run tactics, hunter-killer teams, and small, mobile forces to make it difficult for the Russians to concentrate forces or firepower against them.[149]

Encountering these defenses and intense Chechen resistance, advancing Russian forces "were forced to retreat."[150] The initial attack was an abject failure. Only one unit, the 131st Motor Rifle Brigade (MRB), which was following the Northern Grouping, enjoyed any initial success. Encountering little resistance, the 131st (which consisted of approximately 1,000 soldiers, 26 tanks, and 120 other armored vehicles[151]) was ordered to continue its advance and occupy the train station in the heart of the city.[152] The unit was not successful in its objectives:

> Upon arriving at the train station . . . the brigade failed to carry out the key tasks of securing the area, encircling the station, and posting guards at strategic elevated positions in nearby multistory buildings. Instead, the 131st was extremely careless, leaving their BMPs and many of their weapons in the square in front of the station while most of the personnel congregated inside the building. As a result they were easy prey for the rebel forces that soon surrounded and attacked them.[153]

The Chechen rebels slaughtered the 131st MRB: "By 3 January 1995, the brigade had lost nearly 800 men [of 1,000]." [154] Additionally, they had lost "20 of 26 tanks, and 102 of 120 BMPs, and 6 of 6 ZSU-23's in the first three day's fighting." [155]

---

[149] Thomas, "The Battle of Grozny," p. 7; Arthur L. Speyer, III, "The Two Sides of Grozny," in Glenn, ed., *Capital Preservation*, p. 84.

[150] Kulikov, "The First Battle of Grozny," in Glenn, ed., *Capital Preservation*, p. 45.

[151] Thomas, "The Battle of Grozny," p. 2.

[152] Kulikov, "The First Battle of Grozny," in Glenn, ed., *Capital Preservation*, p. 45.

[153] Kulikov, "The First Battle of Grozny," Glenn, ed., *Capital Preservation*, p. 45.

[154] Thomas, "The Battle of Grozny," p. 2.

[155] Lester M. Grau and Timothy Thomas, "Russian Lessons Learned from the Battles For Grozny," Marine Corps Gazette, Vol. 84, No. 4, April 2000, pp. 45–48.

Quite simply, the initial Russian assault on Grozny failed. With their

> trained force, better tactics, and the advantages of the defense, [the Chechens] were initially able to defeat the poorly trained, undermanned Russian force that sought to capture Grozny without an effective plan.[156]

In the aftermath of their initial failure to take Grozny, the Russians adapted. They brought in reinforcements that

> included elite airborne and *Spetsnaz* troops as well as naval infantry who deployed as complete units—in contrast to the hastily assembled groups that had gone into battle on New Year's Eve.[157]

The Russians also changed their tactics:

> Russian troops learned to methodically capture multistory buildings and defend them. They began to task organize forces into small mobile assault groups, made better use of snipers and heavy artillery, and made sure that units talked to each other and to air assets, so that mutual support was possible.[158]

On January 19, the Russians took the Presidential Palace; on February 22, they were finally able to "seal off the city from the rest of the republic."[159] By early March the rebels had abandoned Grozny and the Russian Federation Ministry of Defense (MoD) forces handed administration of the city over to Russian Federation Ministry of Internal

---

[156] Olga Oliker, *Russia's Chechen Wars, 1994–2000: Lessons from Urban Combat*, Santa Monica, Calif.: RAND Corporation, MR-1289-A, 2001, p. 22. Many of the Chechen rebels had served in the Soviet Army and some were combat veterans.

[157] Oliker, *Russia's Chechen Wars, 1994–2000*, p. 23.

[158] Oliker, *Russia's Chechen Wars, 1994–2000*, p. 24.

[159] Rupe, "The Battle of Grozny," p. 21.

Affairs (MVD) forces. MoD forces largely "moved south to fight the war in the mountains."[160]

The First Chechen War ended on August 22, 1996, with a negotiated cease-fire.[161] In 1999, the Russians would again attempt to reassert control over Chechnya. In this second conflict, however, the Russians employed more resources against a smaller rebel force, as shown in Table 3.7.

Chechnya II

The Chechen war that began in 1999 (and is still ongoing), and its battle for Grozny, were not an exact reprise of the conflict of five years before.[162] This second war was spurred by three factors: increasing crime and anarchy within the semi-independent republic, which had repercussions in the rest of Russia; a desire on the part of many in the Russian government and military to "finish what they started and had not been allowed to finish"; and, to some extent, cynical political motivations on the part of Russian leadership. However, the conflict was more immediately instigated by Chechen incursions into adjacent Dagestan (a neighboring republic in Russia's north Caucasus region) on August 7, 1999. The goal of these incursions was to spur rebellion against Russia in Dagestan.

The Dagestanis were disinclined to follow the Chechen lead, and Russian forces moved within days to assist local police units in their efforts to repel the Chechen forces, which had succeeded in capturing nine villages in their initial assault.[163] The Russian force consisted of troops from the North Caucasus Military Region. Reportedly, they

---

[160] Rupe, "The Battle of Grozny," p. 28. MVD forces never fully controlled Grozny; there was fighting going on throughout the war and large forays into the city by rebel forces. See Rupe, "The Battle of Grozny," pp. 29–31 for a discussion of these rebel forays into Grozny.

[161] Rupe, "The Battle of Grozny," p. 31.

[162] Unless otherwise noted, the source is Oliker, *Russia's Chechen Wars, 1994–2000.*

[163] Stratfor.com, "Recent Military Actions in Border Regions of Chechnya, Dagestan, Georgia Ending," n.d.

Table 3.7
Russian Federal and Chechen Armored Vehicles

Force and Date	Personnel	AFVs	Artillery
Federal Forces Chechnya I (1994–1996)			
December 11, 1994	6,000	500	270
January 1, 1995	8,000	520	340
February 1, 1995	40,000	1,500	397
September 1, 1996	38,000	1,350	350
Rebel Forces Chechnya I (1994–1996)			
December 11, 1994	20,000	134	200
January 1, 1995	40,000	126	190
February 1, 1995	5,000 to 7,000	34	28
September 1, 1996	40,000	48	54
Federal Forces Chechnya II (1999–present)			
December 1, 1999	100,000	1,650	480
Rebel Forces Chechnya II (1999–present)			
December 1, 1999	20,000	14	23

SOURCE: Andrei Korbut, "Learning by Battle," *Nezavisimoye Voyennoye Obozreniye*, December 24, 1999.

also included special mobile battalions specifically trained for mountain fighting.[164]

On August 13, Russian president Vladimir Putin declared Chechen bases to be acceptable targets for Russian attack, and bombing raids into the breakaway region commenced. On August 15, Chechen President Aslan Maskhadov declared a state of emergency in the republic.[165] The Russians made fair progress through Dagestan in August and September. Rebel tactics included surprise attacks, destruction of

---

[164] Igor' Korotchenko, "Brontekhnika Shtormuyet Gori [Armor Storms the Mountains]," *Nezavisimoye Voyennoye Obozreniye*, September 17, 1999.

[165] Stratfor.com, "Recent Military Actions in Border Regions."

LOCs, and efforts to capture key infrastructure. The rebels also sought to control high ground, roads, and towns, and they mined roadways heavily—particularly near towns.[166]

In September 1999 public opinion in Russia, already hostile to the Chechens after five years of Russian propaganda and media coverage of crime and kidnappings, became particularly bellicose after a series of apartment bombings. No individual or group took responsibility, but the bombings were widely attributed by Russian leaders to "Chechen terrorists" (although some evidence suggests that the bombings may have been the work of Russian agents). On October 1, 1999, Russian troops crossed the border into Chechnya and made rapid progress through the northern third of the region, reportedly having it well under control within days.[167]

By the end of October, forces were assembling around the outskirts of Grozny, and bombardment of the city had begun. The lesson that the Russian military had taken from the first Chechen war was that urban combat was more trouble than it was worth. The new approach was to rely on heavy artillery and air bombardment of the blockaded city. (The Russians were so confident in this tactic that they scarcely bothered to train troops for urban combat.) While testing this approach in the towns of northern Chechnya on their way to Grozny, the Russians generally found village elders quick to accept Russian terms. Coercion was one object of this approach, but Russian leaders also believed that the bombing would effectively drive out rebels and destroy whatever enemy defenses and infrastructure were in place. At the same time, Russian forces secured key facilities in the suburbs, where they met with some limited rebel resistance.

Throughout this period, Russian political and military leaders repeatedly issued statements and assurances that they had no intention whatsoever of attacking, or "storming," Grozny. These statements were probably genuine, since the Russians believed that an intensive air and

[166]Oleg Belosludtzev, "Variation in Tactics of Actions," *Nezavisimoye Voyennoye Obozreniye*, May 12, 2000.

[167]Stratfor.com, "Recent Military Actions in Border Regions"; Oliker, *Russia's Chechen Wars, 1994–2000*, p. 39.

artillery barrage would make it possible for a reasonably small force to enter the city and clear it of any remaining resistance. According to reports, Russian planners divided the city into fifteen sectors; in each, they planned reconnaissance operations followed by artillery and air attacks on identified resistance strong points, equipment, and other targets. Corridors would be created for Russian special and loyalist militia forces to advance toward the city center and control key areas. The end result would be a "spiderweb" of Russian control within which motorized rifle troops organized into 30- to 50-man attack groups would eliminate remaining enemy forces, whose mobility would be constrained by the web of Russian control. Final cleanup would be undertaken by Chechen loyalist units.

Early and mid-December witnessed what were probably some reconnaissance-in-force missions (such as fighting in the Khankala suburbs, where Russian forces sought to take control of the airport) and increased calls by Russian officials for civilians to leave the city (safe corridors for departure were promised). Although the government continued to deny that this was an attack on the city, it was clear by December 23 that Russian forces had entered Grozny in significant numbers; 4,000 to 5,000 troops of the total 100,000 deployed to Chechnya had entered the city.

They remained in the city for weeks, fighting a bloody fight that soon made clear that the plans for a Russian spiderweb were wishful thinking. As estimates of enemy resistance remaining in the city climbed, and more and more Russian forces were dispatched to Grozny, it became increasingly clear that Russian claims of their own low losses were inaccurate at best, prevarication at worst. According to Russian sources, the rebels had used underground tunnels, sewers, and bunkers to sit out the bombardment and for resupply throughout the battle. Regardless of how they did it, there were no doubt far more of these rebels than the Russians had expected, and they did not suffer from the blockade to the extent the Russians had planned. While Russian official data do not break casualty counts down into those incurred during the fight for Grozny and those suffered elsewhere, they do record that at least 600 Russians were killed in Argun, Shali, and Grozny between late December 1999 and early January 2000. The losses of individual units tell a different story: the 506th Motorized Rifle Regiment from Privolzhsk lost nearly a quarter of its personnel in the early days of fighting. Other units reported losses of nearly half their men. In the MVD forces, for example, each 50-man company that entered Grozny in December had lost half its men to death or injury by the end of January.

Fighting was brutal and positional, with territory gained and lost repeatedly each day. Time and time again, Russians and rebels edged each other out of the same multistory buildings they each sought to use as strategic positions for snipers. Throughout this battle, the Russians relied heavily on massive firepower, both in the form of artillery and air support. The end result was the destruction of large portions of the city, no doubt with significant civilian casualties.

The main fight ended somewhat unexpectedly, with widespread reports of rebel forces fleeing the city and being destroyed in minefields early on February 2—even though sporadic fighting continued elsewhere in Grozny. Russian authorities initially appeared skeptical of the reports, but it soon became clear that large numbers of rebels had indeed left. The Russians shifted gears, beginning to claim that this was the result of a carefully planned intelligence operation.

Following the withdrawal of large numbers of rebel forces from Grozny in early February 2000, the Russians announced that the fight was now shifting to the mountains (where some level of fighting had been occurring all along).[168] They described this as the final phase of the Chechen military operation and estimated that there were between 5,000 and 7,000 rebels in the mountainous regions. The Russian forces to be sent there included naval infantry (marines), paratroopers, and army forces with experience in mountain fighting. Russian Minister of Defense Igor Sergeev said that the operation's goals were to minimize losses and, if possible, complete the military phase of the operation quickly.

On February 9, Russian authorities announced that major efforts in the mountains would begin two days later, and that forces had

---

[168] Il'ya Maksakov, "Federal'niye Voyska Prodolzhayut Nastupat' [Federal Forces Continue Attack]," *Nezavisimaya Gazeta*, January 13, 2000.

already undertaken some action. In tandem with the announcement, the Russians reportedly bombed mountain villages, in some cases using fuel-air explosives carried by Su-24 aircraft; there were also reports of Russian soldiers looting and pillaging. The Russians admitted the use of fuel-air explosives in the Argun Gorge region, and claimed it was the first—but would not be the last—time such weaponry had been used in this conflict.

On February 11, the Russians banned travel between regions of Chechnya. On February 13, they increased their estimate of enemy force size to 8,000. Border troops with special training, and perhaps some without, were sent to reinforce existing troops in and near the Argun Gorge.[169]

In January 2001 command of the "counterterrorist" operation was officially transferred to the Russian *Federalnaya Sluzhba Bezopasnosti* (FSB) [Federal Security Service], the successor to the KGB. The commander of the joint force grouping on the ground, however, remained Russian Army General-Lieutenant Valery Baranov. Although a range of other units would remain in and deploy to the region, the basis of the force was the Russian Army's 42nd Motor Rifle Division and a brigade of MVD troops.[170]

The Armored Forces

The Russians used a wide array of armored vehicles in the two Chechen Wars, as shown in Table 3.8. Medium-armored vehicles made up the preponderance of the force, with tanks—absent a significant rebel tank threat—serving mainly in a supporting assault-gun role.

---

[169] "Na Argunsoye Ushchel'ye Sbrasivayut Ob'yemno-Detoniruyushchiye Bombi [Fuel-Air Bombs Being Dropped on Argun Region]," Lenta.ru, February 9, 2000; "Osnovniye Boyi V Chechenskikh Gorakh Razvernutsya Cherez Dva Dnya [Main Battles in Chechen Mountains To Begin in Two Days]," Lenta.ru, February 9, 2000; "Russian Force Set to Start Operation in Chechen Mountains," *Jamestown Foundation Monitor*, Vol. VI, No. 29, February 10, 2000; "Rebels Attack 2 Army Trains in Fierce Fight in Chechnya," *New York Times*, February 11, 2000; "V Gorniye Rayoni Chechni Perebrosheno Podkrepleniye [Reinforcements Sent to Mountainous Regions of Chechnya]," Lenta.ru, February 12, 2000.

[170] Mikhail Khodarenok, "Rukovodit' Operatziyey Porucheno Chekistam [Control of Operation Assigned to Chekhists]," *Nezavisimoye Voyennoye Obozreniye* (Internet edition), January 26, 2001.

Employment

The Russians made extensive use of medium-armored vehicles in both Chechen wars. As previously noted, they learned a hard lesson during their initial assault on Grozny during the first war: It was suicidal to lead with armor in a military operations in urban terrain/military operations on urbanized terrain (MOUT) fight. They adapted their tactics, relying more on armor in a supporting role for dismounted infantry and using firepower, and thus eventually took Grozny.

Although the Russians repeated some mistakes of the first war in the second, one clear improvement the second time around was in the use of armor. This included a similar mix of tanks and medium-armored vehicles as in the first war, with MoD tanks supporting the more limited armor of MVD and police forces. Most MVD units had no organic artillery or armor. Moreover, they were not designed or organized for large-scale operations and, prior to the lead-up to the 1999 Chechen war, did not regularly train with units from the armed forces. MoD tanks were, however, more safely and better-employed during Chechnya II. Reactive armor was used more consistently.

Instead of sending armored columns through narrow city streets, as had been done initially in the first war, armor was this time employed to support dismounted infantry units as they made their way through the city. This meant that infantry units protected armored columns as the armor engaged enemy snipers and automatic riflemen in buildings that infantry fire could not reach. Forces were under orders to avoid close combat. On the micro and macro levels, this translated into heavy artillery use. Ground troops probed into new areas to draw Chechen fire, thus exposing enemy positions. Ground troops, retreating to safety, then called in artillery or air strikes on those positions while BMPs with AGS-17 AGLs were used for fire support and to evacuate the wounded.

Russian forces used medium armor in mountain fighting during both Chechen wars, although they also relied heavily on helicopters to deliver men and supplies. The mountains proved challenging in several ways. During the fall 1999 fighting in mountainous Dagestan, Russian military leaders had argued that armor was the most effective means of capturing and holding territory. This doctrine—use armor where

Table 3.8
Russian Armored Vehicles in Chechnya

Type	Weight (tons)	Armament	Max Armor (mm)
T-72 tank	48.95	125-mm gun; 12.7-mm machine gun; 7.62-mm machine gun	N/A[a]
T-80 tank	46.75	125-mm gun; 12.7-mm machine gun; 7.62-mm machine gun	N/A[a]
PT-76 reconnaissance vehicle	16.06	7.62-mm gun; 7.62-mm machine gun	14.0
BMD-1 airborne combat vehicle	8.50	73-mm gun; three 7.62-mm machine guns	23.0
BMD-2 airborne combat vehicle	11.50	30-mm cannon; two 7.62-mm machine guns; Spandrel antitank guided weapon (ATGW) launcher	10.0
BMD-3 airborne combat vehicle	14.50	30-mm cannon; 7.62-mm machine gun; 5.45-mm machine gun; 40-mm grenade launcher; AT-5 ATGW launcher	N/A[a]
BMP-1 IFV	14.80	73-mm gun; 7.62-mm machine gun; Spandrel ATGW launcher	33.0
BMP-2 IFV	15.70	30-mm cannon; 7.62-mm machine gun; ATGM	N/A[a]
BTR-70 APC	12.65	14.5-mm machine gun; 7.62-mm machine gun variants with AGS-17 30-mm AGL or machine guns	9.0
BTR-80 APC	14.96	14.5-mm machine gun; 7.62-mm machine gun	9.0
ZSU-23-4 antiaircraft system	21.00	Four 23-mm cannons	9.2
2S6 antiaircraft system	37.00	Four 30-mm cannons; eight SA-19 antiaircraft missiles	N/A[a]

SOURCES: Foss, *Jane's Tanks and Combat Vehicles Recognition Guide*, pp. 62–63, 66–69, 76–77, 172–179, 410–417; Federation of American Scientists, "Military Analysis Network—2S6M Tunguska Anti-Aircraft Artillery," June 19, 1999; Federation of American Scientists, "Military Analysis Network—ZSU-23-4 Shilka 23MM Antiaircraft Gun," January 22, 1999.

[a] Classified.

it can reach, and where it cannot, move in with dismounted infantry—came from Afghanistan. As a result, significant armored forces (in the form of BTR-70s, BTR-80s, BMP-1s, BMP-2s, BMD-3s, and a "negligible" number of T-72s) were deployed to Dagestan. They supported motor rifle units, delivered infantry to the field of battle, and engaged enemy forces. Commanders described armor losses in Dagestan as "negligible," giving some credit to "built-in" reactive armor on all tanks. This built-in armor was not used in the 1994–1996 Chechen war, where attachable armor, which involved assembly from 240 parts, was notoriously employed.[171] In Chechnya, the Russians found that heavy Chechen mining of roads and passages ensured very slow movement of troops as they sought to clear the mines.[172] Moreover, armored vehicles performed poorly in the adverse terrain and poor weather conditions, getting stuck in mud, snow, and ice, and were vulnerable to a wide range of rebel weaponry.[173]

Key Insights

The two Chechen wars offer several insights about the use of medium-armored forces, designed for conventional warfare and adapted to different roles, in a protracted counterinsurgency that included urban and mountain operations:

- Medium-armored (and heavy) platforms designed for conventional vehicle-on-vehicle combat experienced difficulty operating in other roles. Designed to provide protection in a head-to-head fight, they generally proved vulnerable to top, side, and rear attack by ATGMs, RPGs, and even heavy machine guns.

---

[171] General-Colonel Sergey Mayev, head of weapons and military technology utilization for the armed forces of the Russian Federation, and head of the Main Armor Directorate of the Ministry of Defense, quoted in Korotchenko, "Brontekhnika Shtormuyet Gori [Armor Storms the Mountains]."

[172] The Russian approach to mine clearing was slow and complex. See Lester W. Grau, "Mine Warfare and Counterinsurgency: The Russian View," *Engineer*, Vol. 29, No. 1, March 1999, pp. 2–6.

[173] Roman Boikov, "Luchshe Gor Mogut Bit'... [Better Than Mountains Could Be...]," *Krasnaya Zvezda*, March 17, 2000.

- Weapon systems on fighting vehicles often displayed limited ability to depress and hyperelevate their main guns and coaxial machine guns. This capability was needed to engage targets in multistory buildings or in mountains.
- As a result of their experiences in Chechnya I, the Russians began improving the survivability in their combat vehicles, even to the further detriment of maneuverability and mobility.[174]
- As was the case in Afghanistan, medium armor seemed more sustainable and useful in mountainous terrain than heavy armor, and it provided protected mobility to Russian soldiers.
- Competent infantry who dismount from their vehicles and operate effectively with armored and fire support systems proved crucial in MOUT and mountain fighting.

U.S. Stryker Brigade Combat Teams in Operation Iraqi Freedom (2003–2005)

On March 21, 2003, the United States and its coalition partners invaded Iraq to remove President Saddam Hussein's Ba'athist regime.[175] The initial phase of combat operations featured a rapid advance up the Tigris and Euphrates river valleys, a short bout of intense urban fighting in Baghdad, and the collapse of formal resistance on April 10.[176] President George W. Bush officially announced the conclusion of major combat operations on May 1.[177]

---

[174] R. M. Ogorkiewicz, "Achzarit: A Radically Different Armoured Infantry Vehicle," *Jane's International Defense Review*, September 1995, pp. 73–77.

[175] This discussion addresses U.S. Stryker operations in Iraq between November 2003 and August 2005. Ongoing operations are excluded to preserve operations security (OPSEC). Stryker operations with U.S. special operations forces in Afghanistan and U.S. Air Force units in Iraq are also excluded.

[176] Gregory Fontenot, E. J. Degen, and David Tohn, *On Point: The United States Army in Operation Iraqi Freedom*, Annapolis, Md.: Naval Institute Press, 2005.

[177] U.S. Department of State, "President Bush Announces Combat Operations in Iraq Have Ended," press release, May 1, 2003.

The collapse of the Ba'athist regime created a power vacuum that U.S. and coalition forces were unprepared to fill. The immediate result was an epidemic of disorder and looting, followed by the proliferation of irregular paramilitary organizations ranging from ex-Ba'athist irredentists to Shiite militias, foreign terrorist cells, criminal syndicates, and many others. Though the long-term objectives of these groups were often irreconcilable, they shared a mutual interest in stymieing the occupation. U.S. and coalition forces were soon under heavy pressure from a bewildering array of irregular adversaries who operated separately and in concert, many with support from outside Iraq.

In autumn 2003 the U.S. Army deployed additional forces to counter the increasingly serious irregular threat. Among these forces was the 3rd Brigade of the 2nd Infantry Division (known as "3/2"), the first of the U.S. Army's SBCTs. The 3/2 SBCT operated for a year in northern Iraq and was replaced in autumn 2004 by the 1st Brigade of the 25th Infantry Division (known as "1/25"), another SBCT. The 1/25 SBCT was, in turn, replaced in autumn 2005 by the 172nd SBCT. The 172nd remained in Iraq through spring 2006. To safeguard OPSEC, this discussion is limited to SBCT operations in Iraq between November 2003 and August 2005, encompassing the experience of the 3/2 and 1/25 SBCTs.

The Stryker Brigades

The SBCTs are a recent addition to the U.S. Army's force structure. On October 12, 1999, then–Chief of Staff General Eric Shinseki announced a major initiative to improve the U.S. Army's "strategic responsiveness" by developing a medium-weight brigade equipped with wheeled armored vehicles. He declared that an experimental unit would be established at Fort Lewis, Washington, to evaluate concepts and equipment for the new type of brigade. The unit would initially be equipped with a mix of "off-the-shelf" and prototype vehicles and other systems.[178]

---

[178] Scott R. Gourley, "Stryker Scores with US Tactical Vehicle Force," *Jane's International Defence Review*, June 1, 2006.

The U.S. Army moved quickly to fulfill General Shinseki's vision. The U.S. Army Materiel Command abridged its standard acquisition process, hosting an industry day barely six weeks after General Shinseki's announcement and conducting an "Initial Platform Performance Demonstration" in January 2000. It released a request for proposals a few weeks later and selected the General Dynamics LAV III as the base vehicle for the brigade in November 2000.[179] In February 2002 the vehicle was officially named the "Stryker" and the new brigades were designated "Stryker Brigade Combat Teams."[180] Production Strykers began arriving at Fort Lewis in March 2002 and the first SBCT was certified ready for combat operations in May 2003.[181]

The SBCT is a relatively large brigade composed of three Stryker battalions, a reconnaissance, surveillance, and target acquisition (RSTA) squadron, a field artillery battalion, a brigade support battalion, and company-sized elements of engineers, military intelligence, antitank, and signal troops. Approximately 4,000 troops strong, the brigade possesses 300 more riflemen than a light infantry brigade and twice the riflemen of a mechanized infantry brigade. It also possesses a unique suite of organic capabilities, ranging from unmanned aerial vehicles to SIGINT collection systems, Q-36 and Q-37 counterbattery radars, ground surveillance radars, and tactical human intelligence teams.[182] The SBCT is also equipped with the Army Battlefield Command System (ABCS) and the Force XXI Battle Command Brigade and Below (FBCB2) system, which enable the unit to plan and conduct operations using integrated databases and software planning

---

[179] Gourley, "Stryker Scores with US Tactical Vehicle Force."

[180] U.S. Department of the Army, "Army Announces Name for Interim Armored Vehicle," press release, February 27, 2002. The Strykers are named for two recipients of the Medal of Honor. Private First Class Stuart S. Stryker served with the 17th Airborne Division in World War II. Specialist Robert F. Stryker served with the 1st Infantry Division in Vietnam.

[181] Jack Reiff, "Brigade Combat Team Program Update," briefing, 7th International Artillery and Indirect Fire Symposium and Exhibition, March 21, 2002.

[182] "3-2 SBCT Arrowhead Brigade Capabilities Overview," briefing, February 2006.

tools, real-time situational awareness, and digital messaging down to the vehicle level.[183]

The heart of the SBCT is, however, its Stryker vehicles. More than 300 are fielded in each brigade in nine variants—the Infantry Carrier Vehicle, Anti-Tank Guided Missile, Mortar Carrier, Commander's Vehicle, Medical Evacuation Vehicle, Fire Support Vehicle, Recon Vehicle, Engineer Squad Vehicle, the Nuclear-Biological-Chemical Recon Vehicle, and a cannon-armed Mobile Gun System.[184]

The Stryker is an eight-wheeled LAV. Depending on the variant, it weighs slightly more or less than 20 tons and stands roughly 10 ft tall, 10 ft wide, and 24 ft long. A 350-HP engine provides a top speed of 60 mi/h over an unrefueled range of 330 mi.[185] Armament varies according to vehicle type (see Table 3.9).

On the Infantry Carrier, Commander's Vehicle, Recon Vehicle, and Engineer Squad Vehicle, an M2 .50-caliber heavy machine gun or Mk19 40-mm grenade launcher is mounted in a remote weapon station (RWS) on the roof of the hull. The RWS also contains an imaging sensor, a thermal sensor, and a fire control system that allow the weapon to be employed from within the vehicle. The weapon operator can remain under armor protection, sighting the weapon through a flat-panel display in the hull. However, the RWS is not stabilized. On the other Stryker variants, the weapons are mounted in a traditional cupola, requiring the operator to be partially exposed to employ the weapon.[186]

While the details of the Stryker's armor package are classified, public reports suggest that its basic armor is capable of defeating 14.5-mm machine gun ammunition.[187] According to DoD documents

---

[183] See Daniel Gonzales, Michael Johnson, Jimmie McEver, Dennis Leedom, Gina Kingston, and Michael S. Tseng, *Network-Centric Operations Case Study: The Stryker Brigade Combat Team*, Santa Monica, Calif.: RAND Corporation, MG-267-1-OSD, 2005, passim.

[184] Association of the U.S. Army, *Army Green Book 2005*, Washington, D.C., 2005, p. 360.

[185] *Jane's Armour and Artillery*, 2006, online edition.

[186] *Jane's Armour and Artillery*.

[187] U.S. Government Accountability Office, *Fielding of Army's Stryker Vehicles Is Well Under Way, but Expectations for Their Transportability by C-130 Aircraft Need to Be Clarified*,

Table 3.9
Stryker Variants

System	Weight (tons)	Armament
Infantry Carrier Vehicle	19.0[a]	.50-caliber machine gun or 40-mm AGL
Anti-Tank Guided Missile	20.7	147-mm ATGM; .50-caliber machine gun
Mortar Carrier	21.3	60-mm, 81-mm, or 120-mm mortar; .50-caliber machine gun
Commander's Vehicle	19.1	.50-caliber machine gun or 40-mm AGL
Medical Evacuation Vehicle	18.8	None
Fire Support Vehicle	19.0[a]	.50-caliber machine gun
Recon Vehicle	19.0[a]	.50-caliber machine gun or 40-mm AGL
Engineer Squad Vehicle	20.9	.50-caliber machine gun or 40-mm AGL
NBC Recon Vehicle	19.0[a]	.50-caliber machine gun
Mobile Gun System (2008)	19.0[a]	105-mm cannon

SOURCE: *Jane's Armour and Artillery.*

[a] Weights are "targets" according to Army documents.

available to the public, the Russian KPVT 14.5-mm machine gun is capable of penetrating 30 mm of rolled homogeneous steel armor at right angles at a range of 500 m. This suggests that the Stryker's armor is, to a rough approximation, equivalent to or better than 30 mm of rolled homogenous armor.[188]

Notably, the U.S. Army has chosen to enhance the survivability of the Strykers in Iraq by adding so-called slat armor to each vehicle. The slat system, which weighs approximately 2.5 tons, resembles a large steel cage and is attached to the exterior of the vehicle by mounts that

GAO-04-925, Washington D.C., August, 2004, p. 13.

[188] U.S. Department of the Army, *OPFOR Worldwide Equipment Guide*, U.S. Army Training and Doctrine Command, Deputy Chief of Staff for Intelligence, Fort Monroe, Va., 2001, p. 2-6.1.

position the cage roughly 18 inches from the vehicle's skin.[189] RPGs flying toward the vehicle strike the slat armor first, detonating the RPG warhead and disrupting the formation of the molten jet that is the warhead's primary means of defeating armor. The resulting strike against the Stryker's skin is much less powerful. Anecdotal evidence suggests that the slat system is very, though not totally, effective in preventing RPGs from penetrating the Stryker.[190]

Employment

November 2003 marked the first combat deployment for the Stryker. On its arrival in Iraq, the 3/2 SBCT was immediately committed to Operation Arrowhead Blizzard, a major offensive in and around the city of Samara led by the 4th Infantry Division (known as "4ID"). The 3/2 SBCT conducted cordon and searches, raids, route security, facility security, and a variety of other missions. After two months of intense operations in support of 4ID, the 3/2 SBCT moved north and relieved the 101st Airborne Division (Air Assault) in January 2004.[191]

The Strykers were charged with an extraordinarily difficult mission in northern Iraq. The brigade area of operations (AO) spanned more than 48,000 km^2. Its most important city was Mosul, a large and strategically important urban area with an ethnically mixed population of 1.7 million.[192] As a result, the 3/2 SBCT could only achieve force-to-space and force-to-population ratios that were much less than those widely judged necessary for successful stability operations.[193] The 101st Airborne had previously exerted a measure of control over this

---

[189] Grace Jean, "Stryker Units Win Over Skeptics," *National Defense*, Vol. XC, No. 623, October 2005, pp. 30–35.

[190] U.S. Department of State, "Terrorist Forces Routed in Mosul Region of Iraq, Colonel Says," press release, September 14, 2005.

[191] Steven Sliwa, "Maneuver and Other Missions in OIF: 1-37 FA, 3/2 SBCT," *Field Artillery*, No. PB6-05-2, March–April 2005, p. 12.

[192] Jean, "Stryker Units Win Over Skeptics."

[193] The "20 troops per thousand civilians" rule of thumb first enunciated by James Quinlivan suggests that a force on the order of 34,000 would be necessary to pacify Mosul alone, leaving aside the rest of the brigade's AO. Of course, such rules of thumb are not determinative, but they are suggestive of the challenges facing the SBCTs in Iraq. See James Quinlivan,

area with 25,000 soldiers and two aviation brigades. The 3/2 SBCT was expected to do the same with a much smaller force, in addition to training and advising nascent Iraqi government forces. The 1/25 SBCT and the 172nd SBCT were subsequently assigned to this same sector during their respective rotations.[194]

From a tactical and operational perspective, most Stryker operations in Iraq focused on pacification, the difficult and repetitive task of securing the population and pursuing irregulars and criminals.[195] The SBCTs accordingly operated in small elements throughout the AO, most often at the company level or below. SBCT field artillery battalions were converted into maneuver units to maximize presence.[196] Most day-to-day operations involved patrolling, exerting presence in as much of the area as possible, reacting to contact or incidents, and developing the intelligence necessary to conduct raids and other offensive operations.[197]

Given the difficulty of the assigned mission, the SBCTs acquitted themselves quite well from 2003 to 2005. They disrupted enemy operations in their AO and provided key support to Iraqi forces and governmental authorities. It is still much too early to judge the ultimate effectiveness of the SBCTs, but they do appear to have been at least as effective as other types of U.S. Army units between 2003 and 2005 in the context of their AO in northern Iraq.[198]

From 2003 to 2005, the primary threats confronting the SBCTs in Iraq were IEDs, sniping, RPGs, and mortars. Against these threats, the Stryker's limited armor has proven a less significant handicap than

---

"Burden of Victory: The Painful Arithmetic of Stability Operations," *RAND Review*, Vol. 27, No. 2, Summer 2003, pp. 28–29.

[194] Sliwa, "Maneuver and Other Missions in OIF: 1-37 FA, 3/2 SBCT," p. 12.

[195] Jeff Charlston, "The Evolution of the Stryker Brigade–From Doctrine to Battlefield Operations in Iraq," in John J. McGrath, ed., *An Army at War: Change in the Midst of Conflict*, Fort Leavenworth, Kan: Combat Studies Institute Press, 2006, pp. 48–54.

[196] Sliwa, "Maneuver and Other Missions in OIF: 1-37 FA, 3/2 SBCT," p. 13.

[197] Gourley, "Stryker Scores with US Tactical Vehicle Force."

[198] U.S. Department of State, "Terrorist Forces Routed in Mosul Region of Iraq, Colonel Says."

many anticipated. Roughly 20 Stryker vehicles had been destroyed in combat as of February 2006, out of approximately 1,000 U.S. Army vehicle combat losses in Iraq.[199] The 3/2 and 1/25 SBCTs appear to have suffered 55 soldiers killed between November 2003 and August 2005.[200] The SBCT loss rates therefore appear to be broadly comparable to those of other unit types.

Key Insights

The capabilities and doctrine of the SBCT were a relatively good fit for irregular warfare operations in Iraq between 2003 and 2005. Their use provides the following insights about medium-weight armored vehicles and evolving C2 capabilities that deserve consideration as the U.S. Army continues to deploy SBCTs and develop the FCS:

- The Strykers were sufficiently lethal and agile to operate effectively against irregular adversaries.
- The speed of the Stryker on roads afforded it an ability to react rapidly over relatively extended distances. It was able to do this with a greater level of protected mobility than heavy or light forces.
- The Strykers were sufficiently survivable, with the addition of slat armor, to operate effectively in Iraq between 2003 and 2005.
- The SBCT's digital battle command capabilities provided a useful degree of flexibility and precision in execution. However, they could not provide a picture of the adversary and therefore could not substitute for armor in calculations of survivability.

---

[199] Greg Grant, "Army 'Reset' Bill Hits $9 Billion: Nearly 1,000 Vehicles Lost in Combat," *Army Times*, February 20, 2006, p. 16. According to Grant, by February 2006 U.S. Army vehicle losses included 20 M1 Abrams, 20 Strykers, 50 Bradleys, 20 M113s, 250 HMMWVs, and roughly 650 other trucks, mine-clearing vehicles, and Fox reconnaissance vehicles. Grant also says that U.S. Army helicopter losses in Afghanistan and Iraq included 27 Apaches, 21 Blackhawks, 14 CH-47s, and 23 OH-58s.

[200] See U.S. Department of State, "Terrorist Forces Routed in Mosul Region of Iraq, Colonel Says"; and Adam Lynn, "Back to Iraq, with Tears and Courage," *News Tribune* (Tacoma, Wash.), June 3, 2006.

CHAPTER FOUR

Medium-Armored Forces in Operations at the Lower End of the Range of Military Operations

This chapter examines the employment of medium-armored forces in operations intended for purposes other than combat. The two cases examined are the United States in Task Force Ranger in Somalia (1993) and Australia and New Zealand in East Timor (1999–2000).

The Rescue of Task Force Ranger (Somalia, 1993)

U.S. involvement in Somalia began in August 1992, when CENTCOM began Operation Provide Relief to deliver aid to Somalia. The country had been without a central government since the departure of Mohamed Sirad Barre in January 1991 and was a failed state riven by interclan warfare.[1] UN efforts to aid Somalia had begun earlier, with the passage of UN Resolution 751 in April 1992. This resolution resulted in the UN Operation in Somalia (UNOSOM), which placed UN observers and peacekeepers in Mogadishu.[2]

Despite the delivery of some 28,000 metric tons of relief supplies, Provide Relief was a failure. Quite simply, the scale of the famine and the tenuous security environment required a larger effort. By the fall of

---

[1] Jay A. Hines, "Confronting Continuing Challenges: A Brief History of the United States Central Command," n.d., p. 13.

[2] Joseph P. Hoar, "A CINC's Perspective," *Joint Force Quarterly*, No. 2, Autumn 1993, pp. 56–58.

1992, 500,000 Somalis had died and 1,000 more were dying each day.[3] Consequently, the United Nations passed Resolution 794 in December 1992, authorizing military action to establish a secure environment for aid delivery. This new phase, Operation Restore Hope, lasted from December 9, 1992, to May 4, 1993. The United States provided military forces and led the multinational coalition, United Task Force (UNITAF), in the effort to establish a secure environment that would allow humanitarian assistance to be delivered where it was needed.[4]

UNOSOM II, established by UN Security Resolution 814 on March 26, 1993, replaced UNITAF on May 4, 1993. The UNOSOM II mission was much more expansive than earlier efforts in Somalia. Specifically, UNOSOM II forces had the mandate to disarm the Somali clans and engage in "nation-building," as evidenced by the mission given to the approximately 4,500 U.S. forces who supported UNOSOM II:

> When directed, UNOSOM II Force Command conducts military operations to consolidate, expand, and maintain a secure environment for the advancement of humanitarian aid, economic assistance, and political reconciliation in Somalia.[5]

---

[3] Institute for National Strategic Studies, *Strategic Assessment 1996: Instruments of U.S. Power*, Washington, D.C.: National Defense University Press, 1996, p. 135.

[4] Kenneth Allard, *Somalia Operations: Lessons Learned*, Fort McNair, Washington, D.C.: National Defense University Press, 1995, pp. 5–6. Allard provides the mission statement from U.S. Central Command for UNITAF:

> When directed by the NCA [National Command Authorities], USCINCCENT [Commander in Chief, United States Central Command] will conduct joint/combined military operations in Somalia to secure the major air and sea ports, key installations and food distribution points, to provide open and free passage of relief supplies, provide security for convoys and relief organizations, and assist UN/NGOs [nongovernmental organizations] in providing humanitarian relief under U.N. auspices. Upon establishing a secure environment for uninterrupted relief operations, USCINCCENT terminates and transfers relief operations to U.N. peacekeeping forces.

Allard also notes that "UNITAF ultimately involved 38,300 personnel from 21 coalition nations, including 28,000 Americans."

[5] Allard, *Somalia Operations: Lessons Learned*, pp. 6–7.

The UN forces soon came to be seen by Somalis as having taken sides in the internal Somali conflict, and as threatening to "the Mogadishu power base of one clan warlord, Mohammed Aideed."[6] On June 5, 1993, Aideed supporters killed 24 Pakistani soldiers. The UN Security Council responded swiftly, passing a resolution authorizing the "arrest and detention for prosecution, trial and punishment" of those who had perpetrated the ambush.[7] By June 17, the UN had focused on the apprehension of Aideed, authorizing his arrest in UN Resolution 837. Furthermore, UN Special Envoy Jonathan Howe authorized a $25,000 reward for Aideed's capture. Thus,

> [u]nwittingly, the United Nations hinged its success on Aideed's capture, and U.S. and UN troops stopped being peacekeepers and became one of the warring parties.[8]

In the United States, the force that would become known as Task Force Ranger began forming to execute the mission of capturing Aideed. Task Force Ranger "grew to 450 troops consisting of a squadron from Delta Force, a Ranger company, and elements of the Army's special operations aviation unit, Task Force 160."[9] On August 8, 1993, four U.S. soldiers died when their vehicle was destroyed by a command-detonated mine. That same day, Task Force Ranger was ordered to Somalia; it arrived on August 26 and was ready for operations by August 29. Shortly thereafter, Task Force Ranger began a series of raids to capture Aideed and his key henchmen. In the event that Task Force Ranger needed help, elements of the 10th Mountain Division stood by with a ready quick reaction force (QRF) of infantrymen and AH-1F Cobra attack helicopters on strip alert. Colonel Lawrence E. Casper,

---

[6] Allard, *Somalia Operations: Lessons Learned*, p. 7.

[7] Lawrence E. Casper, *Falcon Brigade: Combat and Command in Somalia and Haiti*, Boulder, Colo.: Lynne Rienner Publishers, 2001, p. 31.

[8] Casper, *Falcon Brigade*, p. 31.

[9] Casper, *Falcon Brigade*, p. 32.

the 10th Mountain Division's Aviation Brigade commander, commanded the QRF.[10]

On October 3, Task Force Ranger launched a daylight raid into the heart of Mogadishu and captured 24 Somalis, including two of Aideed's key lieutenants. Somali militia responded rapidly, however, shooting down two Black Hawk helicopters and trapping the raiders in downtown Mogadishu.[11] The Somali militia also erected makeshift barriers and roadblocks to impede rescue efforts. QRFs from Task Force Ranger and the 10th Mountain Division attempted unsuccessfully to break through to the isolated U.S. raiders: The rescuers' vehicles—lightly armored high-mobility multipurpose wheeled vehicles (HMMWVs) and 5-ton trucks, reinforced with sandbags—were no match for Somali RPGs and machine gun fire, and were turned back before they could extract the raiders.[12]

The Armored Forces

The U.S. forces in UNOSOM II did not have armored vehicles. Major General Thomas M. Montgomery, the deputy commander of UNOSOM II and the senior U.S. officer in Somalia, had requested an augmentation task force of M1 tanks and Bradley fighting vehicles from CENTCOM. General Montgomery believed that such a force

> would protect local logistics traffic, long-haul convoy, key installations and the Mogadishu by-pass observation posts. And it would provide critical road-block clearing for vulnerable, thin-skinned vehicles. I would use it in conjunction with other QRF operations only when necessary.[13]

---

[10] Casper, *Falcon Brigade*, pp. 28–37.

[11] Rick Atkinson, "Night of a Thousand Casualties," *Washington Post*, January 31, 1994, p. A1. See also Mark Bowden, *Black Hawk Down: A Story of Modern War*, New York: Atlantic Monthly Press, 1999, and Casper, *Falcon Brigade*, for in-depth treatments of the October 3 raid and the extended firefight that followed. Casper, *Falcon Brigade*, pp. 60–64, presents a chronology of events and map of the raid.

[12] Casper, *Falcon Brigade*, pp. 43–46.

[13] Casper, *Falcon Brigade*, p. 43.

His request was disapproved by Secretary of Defense Leslie Aspin, Jr.[14] Nevertheless, the QRF clearly needed armor support to rescue the trapped Task Force Ranger soldiers who were fighting for their lives in Mogadishu.

Three of the national contingents in UNOSOM II had armored vehicles—the Italians, the Pakistanis, and the Malaysians. The Italian armor was too far away to provide assistance, but the Pakistanis and Malaysians volunteered their forces. The Pakistanis provided M48 tanks and the Malaysians furnished Condor APCs. The characteristics of these vehicles are shown in Table 4.1.

Employment

The ad hoc force that assembled to break through to the beleaguered soldiers of Task Force Ranger consisted of four Pakistani M48 tanks, 28 Malaysian Condor APCs, and an assortment of thin-skinned vehicles, including HMMWVs armed with MK-19 grenade launchers and machine guns.[15] Infantrymen of the 2nd Battalion, 14th Infantry, replaced the Malaysian soldiers in the Condor APCs, leaving only the Malaysian drivers and turret operators to operate the vehicles and their armament. The Pakistani tanks led and provided security to the

Table 4.1
Armored Vehicles Supporting Task Force Ranger

Type	Weight (Tons)	Armament	Max Armor (mm)
Condor APC (German)	13.6	20-mm cannon; 7.62-mmm machine gun	N/A[a]
M48A3 tank	51.9	90-mm gun; .50-caliber machine gun; 7.62-mm machine gun	120

SOURCE: Foss, *Jane's Tanks and Combat Vehicles Recognition Guide*, pp. 116, 260.

[a] Classified.

---

[14] Bolger, *Death Ground*, p. 209.

[15] Bolger, *Death Ground*, pp. 223–224.

convoy on the first leg of the relief mission, but broke off as planned before reaching Task Force Ranger. The Malaysian Condor APCs, supported by U.S. gunships, fought their way to the stranded Task Force Ranger soldiers and evacuated them.

At 6:32 p.m., the convoy reached friendly lines. After some 15 hours of combat, Task Force Ranger's raiding force was finally safe.[16] The Malaysian Condors provided the protected mobility and firepower that made their rescue, after two failed U.S. attempts, possible.

The costs were high. U.S. casualties included 18 dead and 84 wounded. The Malaysians suffered one dead and seven wounded. Two Pakistanis were wounded. In addition to casualties, Task Force Ranger and the relieving forces lost two helicopters and damaged four. The Malaysians lost four Condor APCs to Somali RPG fire, and several U.S. HMMWVs and trucks were destroyed or damaged. Estimates of Somali casualties were much higher: 312 dead and 814 wounded. U.S. policy in Somalia was also a casualty. On October 7, President William J. Clinton announced that all U.S. forces would withdraw from Somalia by March 31, 1994.[17]

Key Insights

The firefight discussed in this case was the most intense that U.S. forces had engaged in since the Vietnam War. Furthermore, the battle in Mogadishu was the first urban combat operation of significant scale for U.S. forces since the Vietnam War. Several key insights can be drawn from this case:

- Even highly trained light and special operations forces—with ready access to massive firepower and helicopter support—could be cut off in a MOUT fight by determined opponents.
- Medium-armored vehicles provided the decisive edge in extracting U.S. light forces in Mogadishu because of their protected mobility and on-board firepower.

---

[16] Casper, *Falcon Brigade*, pp. 43–87.

[17] Casper, *Falcon Brigade*, pp. 85–93.

- Prior coordination (and training and rehearsals, if possible) between medium-armored and light forces is crucial, particularly if the forces hail from different nations.

Australia and New Zealand in East Timor (1999–2000)

In 1975 Indonesia invaded East Timor. In the decades that followed, an active insurgency contested Indonesian rule, and tens of thousands died in the conflict.[18] In January 1999 the potential for an end to violence in East Timor materialized when Indonesia's President B. J. Habibie indicated that "his government might be prepared to consider independence for East Timor."[19] On May 5, 1999,

> three agreements were concluded . . . variously between Indonesia, Portugal and the United Nations, creating a framework for the resolution of East Timor's future status. It was envisaged that the People's Assembly would implement the results of the popular consultation in October 1999.[20]

On May 7, 1999, the UN Security Council adopted Resolution 1236, setting in motion steps to hold a popular referendum to determine the future of East Timor. The citizens of East Timor would vote for or against integration with Indonesia. A vote against integration would be a vote for independence. The UN also established the United Nations Mission in East Timor (UNAMET) to, in the words of UN Special Representative to East Timor Ian Martin, ensure "a fair campaign and ballot."[21]

---

[18] Michael J. Kelly, Timothy L. H. McCormack, Paul Muggleton, and Bruce M. Oswald, "Legal Aspects of Australia's Involvement in the International Force for East Timor," *International Review of the Red Cross*, No. 841, March 31, 2001, pp. 101–139.

[19] "The United Nations and East Timor: A Chronology," United Nations Peace and Security Web site, n.d.

[20] Kelly et al., "Legal Aspects of Australia's Involvement," p. 4.

[21] "The United Nations and East Timor: A Chronology."

On August 30, 1999, despite intermittent violence, the citizens of East Timor voted overwhelmingly in favor of independence from Indonesia.[22] In the aftermath of the referendum, local militia groups, backed by the Indonesian army, went on a rampage and reportedly engaged in "Balkan-style atrocities."[23]

On September 7, in the midst of the mayhem, Indonesia declared martial law in East Timor. The UN Security Council sent a mission to Jakarta and Dili from September 8–12, and reported that the Indonesian military and police "were either unwilling or unable to provide an environment for the peaceful implementation of the 5 May agreements."[24] On September 12, the Indonesian government agreed to international intervention to restore peace in East Timor. On September 15, the UN Security Council adopted Resolution 1264, under Chapter VII of the UN Charter, to deploy a multinational force to East Timor to

> restore peace and security in East Timor, to protect and support UNAMET in carrying out its tasks, and, within force capabilities, to facilitate humanitarian assistance operations, and *authorize* . . . the States participating in the multinational force to take all necessary measure to fulfill this mandate.[25]

The multinational force, called International Force East Timor (INTERFET), began deploying on September 20. The situation they confronted was dire:

> In Dili few buildings were still intact. The towns of Ainaro and Cassa had been completely destroyed, with an estimated 70 per

---

[22] "The United Nations and East Timor: A Chronology." In this election, 344,580 East Timorese (78.5 percent of the population) voted for independence, while 94,388 (21.5 percent) voted for "the special autonomy proposal" to keep East Timor under Indonesian rule.

[23] Craig A. Collier, "A New Way to Wage Peace: US Support to Operation Stabilise," *Military Review*, Vol. 81, No. 1, January–February 2001, pp. 2–9.

[24] Kelly et al., "Legal Aspects of Australia's Involvement," p. 4.

[25] The United Nations Security Council, *Resolution 1264*, S/RES/1264, September 15, 1999.

> cent of Atsabe, Gleno, Lospalus, Maliana, Manatuto and Oecus either burnt down or leveled . . . . The judicial and detention systems were not operating and no commercial activity was being conducted. There was no effective administration, as administrative officials had apparently left the territory after the announcement of the ballot results.
>
> In humanitarian terms, the situation in East Timor when the multinational force arrived there was one of crisis: a preliminary UN interagency assessment, issued on 27 September 1999, estimated that of a total pre-ballot population of 890,000, over 500,000 had been displaced by the violence, including 150,000 to West Timor.[26]

As the multinational force began arriving in East Timor, the Indonesian government ended its state of martial law and withdrew its military forces.[27]

Although 22 nations eventually participated in INTERFET, Australia provided the majority of the "combat/combat support personnel, armoured vehicles and mobility and logistics assets."[28] New Zealand provided a battalion combat group.[29] Together, Australia and

---

[26] Kelly et al., "Legal Aspects of Australia's Involvement," p. 4.

[27] Kelly et al., "Legal Aspects of Australia's Involvement," p. 4.

[28] Ian Bostock, "East Timor: An Operational Evaluation," *Jane's Defense Weekly*, Vol. 33, No. 18, May 3, 2000, p. 24. See also Kelly et al., "Legal Aspects of Australia's Involvement," p. 5, which approximates INTERFET's total force strength at 12,600 troops, of which the Australian Defence Force (ADF) furnished 5,521 troops.

[29] Alan Ryan, *Primary Responsibilities and Primary Risks: Australian Defence Force Participation in the International Force East Timor*, Duntroon, Australia: Land Warfare Studies Centre, 2000, pp. 127–129. Annex A of Ryan's study lists the contributions of the various nations supporting INTERFET. Ryan notes that Australia provided the following forces:

> *Maritime*: 3 x Frigates, 1 x Landing ship, 3 x Landing Craft, 1 x tanker, 1 x Jet Cat, 1 x Clearing Team. *Land*: HQINTERFET, 1 x Joint Support Unit, Brigade Headquarters, 10 Signals Squadron, 2 x Infantry battalion groups, 1 x Mechanized battalion group, Special Forces, 1 x Armoured Personnel Carrier Squadron, 2 x Construction Squadron, 1 x Aviation Regiment, 1 x Reconnaissance Squadron, 1 x Brigade Administrative Support Battalion, 1 x Forward Logistic Support Group, 1 x Forward Support Base, Combat Engineer Regiment. *Air*: 12 x C130, 2 x 707, 4 x Caribou aircraft.

New Zealand "provided four fifths of the early operation capacity of the force."[30]

Major General Peter Cosgrove, Commander of the ADF's Deployable Joint Force Headquarters, commanded INTERFET. Major General Songkitti Jaggabatara served as his deputy.[31] Following a predeployment preparatory phase, INTERFET executed a four-phase operation called Operation Stablise. Before the deployment of INTERFET, however, General Cosgrove visited East Timor and conferred with Major General Kiki Syahnakri, the senior Indonesian officer in East Timor. General Cosgrove recalled the meeting:

> In an initially cautious but positive meeting with my counterpart, Major General KiKi Syahnakrie of TNI [*Tentara Nasional Indonesia* (Indonesian National Army)], we negotiated the details of my initial requirements for airfield and port use and deployment areas. He and his advisers seemed taken aback at the size and rapidity of the initial deployments and my clear intention to embark immediately on stability operations in Dili. I used here what I suppose is best described as a Rooseveltian approach ('Speak softly' etc) which I tried to apply throughout Operation Stabilise.[32]

The next day, Operation Stabilise commenced in accordance with General Cosgrove's four-phased operational plan, described below:

> Phase 1—Control: during this phase, INTERFET control was established over air and sea points of entry in Dili on 20 Sep-

---

According to Ryan, New Zealand contributed the following forces to INTERFET: "*Maritime*: 1 Frigate, 1 Tanker. *Land*: Infantry battalion group. *Air*: 2 x C130, 6 x Helicopters." See also Phil Gibbons, "The Urban Area During Stability Missions Case Study: East Timor," Glenn, ed., *Capital Preservation*, pp. 99–161, for a New Zealand perspective on INTERFET.

[30] Center for Strategic Studies: New Zealand, "Strategic and Military Lessons from East Timor," *Center for Strategic Studies Briefing Papers*, Vol. 2, Part 1, February 2000.

[31] Ryan, *Primary Responsibilities and Primary Risks*, pp. 68–69.

[32] Ryan, *Primary Responsibilities and Primary Risks*, p. 69.

> tember 1999 and an air point of entry in Bacau on 23 September 1999;
>
> Phase 2—Consolidation: this phase occurred in the period September 1999 to January 2000 and involved INTERFET establishing and maintaining control progressively throughout East Timor, including the Oecussi enclave in West Timor and Atauro Island;
>
> Phase 3—Transition: INTERFET objectives were to hand over control of East Timor to UNTAET [UN Transitional Administration East Timor], having maintained security for three months without a serious incident, set up a border security management system, established an internally displaced persons (IDP[s]) return plan and reduced the risk of militia activity. The transition from INTERFET to UNTAET took place progressively from east to west. Sector East was handed over on 1 February 2000, Sector Central, including Dili, on 14 February 2000, the Oecussi enclave on 15 February 2000 and Sector West on 21 February 2000;
>
> Phase 4—Redeployment: INTERFET formally handed over authority to UNTAET on 23 February 2000 with INTERFET troops either moving to the UNTAET command structure or redeploying to home locations.[33]

By almost all measures, INTERFET was a success. Indonesia's decision to withdraw from East Timor, coupled with the inability of the militia to carry out its threat to violently oppose INTERFET, did play a role in this success. Nevertheless, the rapidity with which INTERFET established control also stymied the militia. In the end, INTERFET rapidly created a secure environment in East Timor that enabled the transition rebuilding efforts.[34] Indeed, the mission has been lauded as

---

[33] Kelly et al., "Legal Aspects of Australia's Involvement," pp. 5–6. The overall Australian effort, which included support operations from Australia, was called Operation Warden.

[34] Ryan, *Primary Responsibilities and Primary Risks*, pp. 68–76.

> a benchmark against which similar future peace support operations (PSOs) will be assessed . . . . INTERFET's 'Operation Stabilize' met its military objectives, facilitating favourable outcomes in the political arenas of participating countries and the UN.[35]

Furthermore, the operation resulted in "no killed in action, no own casualties and no collateral damage among the East Timorese."[36]

The Armored Forces

Both Australia and New Zealand deployed medium-armored vehicles with their INTERFET contingents. Australian armor included M113 APCs and Australian light-armored vehicles (ASLAVs); New Zealand deployed M113s.[37] The characteristics of these vehicles are depicted in Table 4.2.

Australian armored vehicles in East Timor were medium-armored when compared to the heavy armored forces (i.e., Leopard AS1 MBTs) in the Australian Army that were not deployed as part of INTERFET.

Employment

The Australian Army deployed ASLAVs and M113s without their Leopard AS1 MBTs. Two factors played into this decision: the absence of enemy armor in East Timor and the imperative to get the force to East Timor rapidly. The commander of the 2nd Cavalry Regiment, Lieutenant Colonel Michael Krause, believed "the army's response in East Timor—weighted in favour of rapid deployability over firepower, protection and close-combat—was an appropriate and measured one."[38]

---

[35] Bostock, "East Timor: An Operational Evaluation," p. 23.

[36] Bostock, "East Timor: An Operational Evaluation," p. 23.

[37] ASLAV is the Australian variant of the General Motors LAV series.

[38] Bostock, "East Timor: An Operational Evaluation," p. 26. Colonel Krause, however, indicated that he would have preferred to have had the Leopards in East Timor:

> [I]f it [the largely unopposed lodgements and subsequent deployment of INTERFET forces] had gone a different way and we didn't have the Leopards, then we would have been fighting with one arm behind our backs. . . . Would I have liked to have seen Leopards in Dili? Of course I would.

Table 4.2
Australian and New Zealand Armored Vehicles in INTERFET

Type	Weight (Tons)	Armament	Max Armor (mm)
M113[a]	12.5	.50-caliber machine gun; 7.62-mm machine gun	38
ASLAV Type I[b]	14.1	25-mm chain gun; two 7.62-mm machine guns	10
ASLAV Type II	14.2	7.62-mm machine gun	10
ASLAV Type III	14.2	7.62-mm machine gun	10

SOURCES: Foss, *Jane's World Armoured Fighting Vehicles*, p. 294; Australian Department of Defence, "LAND 112–ASLAV (Australian Light Armoured Vehicle)," September 21, 2007.

[a] Foss, *Jane's World Armoured Fighting Vehicles*, p. 295. The Australian variant of the M113 mounts the T50 turret with a .50-caliber machine gun and a 7.62-mm machine gun.

[b] See Australian Ministry of Defence, "LAND 112—ASLAV," which notes the following:

> The ASLAV family is comprised of seven mission role variants derived from three similar vehicle types. The Type I vehicle is the ASLAV-25, distinguished by its two-man drop-in turret. The Type II vehicle adapts a variety of roles by the use of non-permanent Mission Role Installation Kits (MRIKs), allowing commanders to reconfigure their force composition as needed. Type III vehicles are similar to Type II, with structural modifications to support the winch and crane mounts, anchor points and winch cable entry required for field maintenance roles.

Type I is the ASLAV 25; Type II includes the ASLAV-PC (Personnel Carrier), ASLAV-S (Surveillance), ASLAV-C (Command), and the ASLAV-A (Ambulance); and Type III includes the ASLAV-R (Recovery) and ASLAV-F (Fitters).

The decision to deploy medium armor only enabled General Cosgrove to rapidly deploy substantial INTERFET forces into Dili. At dawn on September 20, Royal Australian Air Force (RAAF) C-130s brought Australian, New Zealand, and British special forces into Dili's Komoro Airport. Later that day, RAAF C-130s brought in the Second Battalion of the Royal Australia Regiment, two M113A1 APCs, and a British Gurkha company from Brunei. On September 21, the *Jervis Bay* brought the Third Battalion of the Royal Australian regiment to Dili port while the HMAS *Tobruk* delivered 22 ASLAVs from C Squadron,

2nd Cavalry Regiment. This force was joined by 12 Black Hawk helicopters that self-deployed from Australia.[39]

General Cosgrove's initial ground force was largely composed of Australian and New Zealand troops. His strategy was to secure Dili as a base of operations. As INTERFET forces from other nations arrived, he employed the "oil-spot" strategy, "based on the idea that INTERFET, having first secured Dili, could then spread to other strategic locations as more troops arrived in country. From these individual 'oil spots,' ever-larger areas would be brought under UN control."[40]

The medium armor available to the Australian and New Zealand Forces proved very useful in East Timor. M113s and ASLAVS were used in a wide variety of roles. They provided "infantry elements with protected tactical and combat mobility, and conducted escort, security, surveillance, reconnaissance, response, communications, search, vehicle check points and force presence operations."[41] Despite its age, the M113 proved particularly useful:

> The M113's superior cross-country mobility often meant it was the only vehicle type able to deploy or redeploy infantry patrols, sniper teams, civil-military operations teams and retransmission sites to remote villages and border areas. When poor weather in East Timor's high country prevented helicopter operations, the M113 was the only Australian platform capable of fulfilling these key mobility tasks.[42]

---

[39] Ryan, *Primary Responsibilities and Primary Risks*, pp. 68–70.

[40] Ryan, *Primary Responsibilities and Primary Risks*, p. 70.

[41] Bostock, "East Timor: An Operational Evaluation," p. 25, and New Zealand Army, "East Timor," n.d.

[42] Bostock, "East Timor: An Operational Evaluation," p. 26. See also Michael Rose, "A Liddell Hart Approach to Peacekeeping," Liddell Hart Centre for Military Archives, King's College, London, 1999. General Sir Michael Rose, commander of the United Nations Protection Force in Bosnia, supports General Cosgrove's view that peacekeeping forces should be trained as warfighters:

> I firmly believe that in all circumstances a force must be organised and equipped as a war fighting force, able to conduct peace enforcement operations from its very first moment of deployment, commensurate with the highest point of the spectrum to which one must reasonably expect to have to escalate in a 'worst case scenario.'

Key Insights

INTERFET operations during Operation Stabilise offered the following key insights about the utility of rapidly deployable medium-armored forces in peacekeeping operations and about the relative advantages of wheeled versus tracked vehicles:

- Medium armor gave INTERFET protected mobility, mobile firepower, and a means of intimidating the militia that light infantry could not provide.
- Tracked vehicles had a significant off-road mobility advantage over wheeled vehicles in East Timor.
- Medium armor enabled the Australian and New Zealand contingents to rapidly deploy with armored protection, mobility, and lethality. The low threat level facilitated the decision to leave Leopard MBTs behind, although they remained ready to deploy if the situation in East Timor worsened.[43]

[43] Bostock, "East Timor: An Operational Evaluation," p. 26.

CHAPTER FIVE

Conclusions

The case studies examined in this monograph show that forces equipped with medium-armored platforms have been employed across the range of military operations and in virtually every type of complex terrain. This final chapter returns to the three questions that guided the study and addresses each in turn. Finally, it presents overarching insights about medium-armored vehicles and discusses implications for the U.S. Army's Future Force and the FCS.

What Unique Capabilities Have Medium-Armored Forces Brought to Past Conflicts, and Where Along the Spectrum of Operations Have They Been Most Valuable?

Forces composed of medium-armored vehicles have shown unique capabilities and utility across the range of military operations since the inception of mechanization. As demonstrated in the cases examined in this monograph, medium-armored vehicles have been particularly valuable in the middle and at the lower end of the range of military operations. Several cases show the critical difference that even small numbers of medium-armored forces can make. In Somalia, Malaysian medium armor provided the protected mobility and firepower necessary to extricate cutoff elements of Task Force Ranger. U.S. mechanized forces in Panama provided a needed edge to light forces, and even the modest number of deployed M551 Sheridans provided an important capability at crucial moments in the early stages of the campaign. Medium-armored forces from Australia and New Zealand gave INTERFET the capability to range widely and rapidly across East Timor with protected mobility and lethality sufficient to meet the threat. Soviet medium-armored platforms provided sufficient armor protection, mobility, and lethality for Soviet coup de main operations in Czechoslovakia and Afghanistan. Finally, SBCTs were able to provide rapid response across a large operational area in Iraq, providing greater survivability than light forces. In each of these cases, medium-armored vehicles allowed forces to perform better than light forces alone by providing protected mobility, mobile firepower, and a rapid reaction capability that footsoldiers or truck-borne infantry do not possess.

Enemy capabilities—for example, a lack of tanks or operational competence—rendered heavy-armored vehicles unnecessary in many cases (Vietnam, Czechoslovakia, Afghanistan, Panama, Somalia, Chechnya, East Timor, Desert Storm, and OIF). In Angola, SAA forces were able to operate effectively against MPLA tank-equipped forces. The South African advantage in training and situational awareness made the difference for some time. Nevertheless, the South Africans eventually decided to field their own tanks to overcome the inherent lethality and survivability problems of their medium-armored vehicles.

Absent a threat of tank-on-tank combat, medium armor was more effective in many situations than heavy armor would have been. Medium-armored vehicles generally were able to move more rapidly over the operational area, particularly in environments with complex terrain and deteriorated infrastructure. Furthermore, they usually required less logistical support than heavy armor. Finally, the weapons on medium-armored vehicles were more readily adaptable to combat in mountains (Afghanistan, Chechnya) and urban areas (Chechnya) than heavy armor. This is because the main guns on tanks generally cannot hyperelevate or depress for all targets. Tanks are also equipped with relatively low-caliber machine guns that do not accommodate explosive ammunition, and their coaxial machine guns demonstrate the same elevation problems as their main guns. Providing hyperelevating, rapid-firing, medium-caliber weapons with explosive ammunition for medium-armored vehicles—rather than modifying tanks—was the approach taken to solve this problem in most cases.

At the higher end of the range of military operations—when the opponent had heavy armor—the story of the effectiveness of medium armor is mixed. During Operation Desert Storm, U.S. Marine Corps LAV-equipped units took advantage of their superior training and fire-support resources to defeat a much less-competent Iraqi force equipped with tanks. In Spain, however, Russian tanks had a clear advantage over more lightly armed German and Italian vehicles. In World War II, U.S. tanks and tank destroyers operated with an enormous lethality and survivability disadvantage against competent German forces equipped with tanks and antitank weapons. The U.S. Army compensated for the shortcomings of their medium-armored vehicles by overwhelming the Germans with sheer numbers and very effective artillery and air support.

One additional insight in this area deserves attention from the U.S. Army as it prepares to field its future forces: The exquisite situational awareness the United States enjoys in conventional force-on-force conflict is not always sufficient to meet the demands of irregular warfare. Here, the case of the SBCTs in Iraq between 2003 and 2005 is particularly important because it demonstrates that even excellent situational awareness is not a substitute for traditional armor. The SBCTs possessed the most advanced digital battle command systems available to any army at the time, and through these systems the SBCT could access national-level sources of intelligence and information. Even with these capabilities, however, there is no evidence that the SBCT was able to significantly enhance its survivability through detailed situational awareness of the enemy's capabilities, intentions, and dispositions. The nature of the irregular adversary rendered much of this capability inapplicable. This suggests that while digital battle command systems are very useful, particularly in making planning and execution more rapid and precise, they cannot substitute for armor protection.

How Have Medium-Armored Forces Performed in Complex Terrain in the Past?

It is important to note that in almost every case examined in this monograph, medium-armored forces had to operate in some form of "close" complex terrain, i.e., urban, jungle, or mountainous, or some combination thereof. Furthermore, most of the operations were affected by the underdeveloped infrastructure that characterized the operational environments. These factors generally place higher demands on CSS functions because of heavier maintenance demands on equipment and the absence of host nation support. And in the case of several protracted conflicts (Afghanistan, Chechnya, OIF), the infrastructure continued to deteriorate as the war continued. Finally, medium-armored forces are more able to operate in areas with less-developed infrastructure. This was the case in Panama, where M551 Sheridans could cross bridges that would not support U.S. MBTs.

Naturally, complex terrain frequently restrained the tempo of maneuver. There is, however, another important dimension of complex terrain that is evident in almost all of the cases examined in this monograph: Complex terrain imparts an advantage to the defender and often creates survivability problems for the attacker. This is true for two reasons.

First, until very recently, most armored vehicles, including medium-armored platforms, were designed for employment in conventional combat. In this kind of warfare, the expectation is that armored vehicles face the greatest threat from other armored vehicles in head-to-head direct-fire engagements. Therefore, the armor is thickest on the front of combat vehicles. In complex terrain, however, direct-fire attacks frequently occur at close range and are aimed at the more vulnerable sides, rear, or tops of vehicles. Additionally, the belly of most armored vehicles is thin and thus vulnerable to mines and IEDs, weapons that are easier to conceal in complex terrain. Furthermore, complex terrain marginalizes the lethality advantage usually afforded by standoff fires and degrades the ability to see first.

Second, in the urban canyons that characterize many large cities, the ability to employ weapons from protected firing positions (and at very high angles of elevation) is important. This was particularly apparent in Grozny, where Russian forces eventually had to deploy air defense guns to be able to engage Chechen fighters in the upper stories of buildings. It was also an important consideration in the mountainous terrain that characterized much of the Soviet fighting in Afghanistan. As previously noted, operations in urban and mountainous terrain require weapons of medium caliber (20 to 35 mm) that can be fired on the move, offer high rates of accurate fire, and accommodate explosive ammunition; these characteristics are particularly important in ambush situations. Furthermore, gunners should ideally be able to fire weapons from under cover to avoid exposure to sniper fire and fragmentation from IEDs, RPGs, and other weapons. In the cases examined in this study, medium-armored vehicles were better than tanks at operating with mobility in complex terrain and accommodating the needed weapons. Medium-armored platforms also provided greater survivability to infantry than light vehicles did.

What Advantages Has the Rapid-Deployment Capability of Medium-Armored Forces Provided to Operational Commanders in the Past?

The capacity to rapidly deploy medium-armored forces may be an important national capability, as was apparent during South Africa's involvement in Angola. It was also an important capability in the coup de main operations conducted by the Soviet Union in Czechoslovakia and Afghanistan. The Soviets chose medium armor for these operations, preferring their greater deployability compared to heavy forces and their greater mobility and firepower compared to light forces. Additionally, the air-dropped Sheridans used in Panama provided an important capability to U.S. light forces, although they were matched against an already vastly out-classed enemy. However, it is important to note that deploying medium-armored forces would not be a sustainable long-term strategy against forces with heavy armor.

Nevertheless, Operation Just Cause reveals an existing U.S. Army capability gap: The U.S. Army lacks a forced-entry, air-droppable medium armor capability. In Panama, the M551 Sheridan provided this capability, but this vehicle has since been retired from the inventory. Stryker medium-armored vehicles are not air-droppable and, with their add-on armor, can only be deployed by C-117 or C-5 transport aircraft. This likely limits their movement by air to any but secure locations.

Implications for the U.S. Army Future Force and the FCS

Three overarching insights with relevance to the U.S. Army as it develops concepts and technologies for the future have emerged from an assessment of these case histories.

Conceptual Choices Matter and Shape Initial Capabilities

The capabilities a nation possesses when a conflict begins or a contingency arises are those that will initially be available to prosecute that conflict. Thus, armies often find themselves in conflicts that differ both from what they have been expecting and preparing for. Similarly, the majority of the difficulties medium-armored forces have experienced in operations have had their origins in conceptual decisions made in peacetime. This phenomenon is most apparent in the case of the U.S. Army in World War II, when U.S. medium-armored forces had to fight German heavy armor that possessed significant survivability and lethality advantages. This was a circumstance that U.S. Army doctrine assumed would not be necessary; this conceptual failure resulted in more U.S. casualties. Similarly, the SAA eventually revisited its assumptions about the viability of medium armor against heavy armor in its war in Angola, eventually fielding its own MBT as a result.

An important message from this monograph is that the U.S. Army, as it develops the Future Force, must employ a broad conceptual framework that embraces the complexity and diversity of the types of military operations that the nation may call upon that force to execute. In future conflict environments, U.S. Army medium-armored forces will likely face—as they have in the past—forces with heavy armor and antitank weapons. They will also need to operate in complex terrain. These factors may make medium-armored forces less survivable than heavy-armored forces. Several cases examined in this monograph (U.S. Marine Corps LAVs in Desert Storm, South Africa in Angola) show that these vulnerabilities can be mitigated by preparing for combined-arms operations at the lowest levels and leveraging organic and external firepower. That said, in complex terrain and irregular warfare, improvements in situational awareness have thus far not proven to be a reliable substitute for armored protection.

The U.S. Army's World War II experience in northwest Europe is particularly instructive. U.S. armored forces were clearly disadvantaged when they confronted better-equipped German armor. This inferiority did not originate in materiel design, however—it stemmed from conceptual decisions, made by the U.S. Army before the war, about how armor should be employed.

The experience of the interwar U.S. Army shows that constraints—like weight restrictions—can impede the realization of important capabilities. In fact, it was a U.S. War Department requirement that limited tanks to a weight of 15 tons that prevented tanks from attaining the operational capabilities required by the infantry branch during World War II. A later requirement that U.S. Army tanks not exceed 30 tons further constrained U.S. tank development. More importantly, a major key conceptual assumption—that proved false—was that U.S. tanks would not fight other tanks. Thus, the U.S. Army doctrine for armored warfare did not include provisions for fighting other armored forces or see a need to integrate air power as the Germans had in their concept for blitzkrieg. But the U.S. Army found itself in a different war than it had envisioned. In consequence, U.S. armored vehicles were markedly less lethal and less survivable than high-end German tanks and tank destroyers, and they suffered as a result. The United States had to rely on its overwhelming numbers and firepower to equalize the fight, and had to invent processes to integrate ground power and air power while under fire.

More recently, the U.S. Army mandated that the interim combat vehicle be transportable by C-130 cargo aircraft. In a decision made partly to help control the vehicle's weight so that this primary requirement could be met, the U.S. Army decided that armor sufficient to protect the vehicle against 14.5-mm machine gun fire would be adequate. In Iraq, where Strykers are now deployed, this level of protection is not sufficient. While slat armor has been added to increase Stryker survivability, it has rendered the vehicle impossible to transport by the C-130.

Quite simply, emerging concepts and desired capabilities—such as rapid deployment by a class of airlift—can place restrictions on developing weapon systems that force critical capability trade-offs. In the case of the Sherman tank and the Stryker vehicle, it was weight constraints that resulted in the construction of vehicles that proved deficient in lethality or survivability in particular combat environments. These outcomes should be considered during the design of the FCS.

Medium Armor Can Be Disadvantaged Against Competent Heavy Forces and Vulnerable in Complex Terrain

When medium-weight forces have been pitted against competent heavy armor—as the U.S. Army was in World War II—they have fared quite poorly. This is relevant for the FCS if the system's survivability and lethality, which are based on what the U.S. Army characterizes as high-risk technologies, do not live up to expectations or cannot be fully employed in some battlefield environments. For example, if the FCS sensor networks are degraded by complex terrain, or if its active protection systems are unusable in close proximity to noncombatants or friendly dismounts, the FCS could encounter a "Sherman dilemma" on a future battlefield. As previously noted, the U.S. M4 Sherman tank was tailored to a very specific tactical concept and key assumptions about the enemies it would fight. The Normandy campaign and subsequent campaigns in northwest Europe did not comport with those assumptions, and the Sherman was clearly outclassed by the German heavy armor and antitank weapons it confronted. Its performance suffered accordingly, and U.S. soldiers paid a heavier price than they should have.

Our analysis suggests that the FCS might encounter its own Sherman dilemma if technological or environmental conditions eliminate its direct overmatch against heavy-threat armor. The U.S. Army would be wise to allocate adequate research, modeling, and experimentation resources to avoid this potential scenario.

Combined Arms and Force Quality Can Mitigate the Inherent Advantages of Heavy Armor

Another major finding derives from the Sherman dilemma: Doctrinal and organizational steps can, in certain circumstances, mitigate the disadvantages that medium-armored vehicles face when confronted by heavy armor. These steps include high-quality combined-arms tactics down to the lowest echelons, effective application of supporting firepower, and excellent all-around training for crews and junior leaders. Some or all of these measures played a role in, for example, South Africa's ability to contend with heavy Angolan and Cuban armor for as long as it did, eventual U.S. triumph in northwest Europe, and the success of Task Force Shepherd during Desert Shield and Desert Storm. The U.S. Army has pursued many of these steps; for example, it has established combined-arms maneuver battalions in the new BCT design, and has for many years provided rigorous training to units at combat training centers. The U.S. Army should continue to explore and experiment with these and other measures that could further mitigate any risk associated with FCS technology or the environments in which the Future Force may fight. In sum, nontechnical steps may be necessary to mitigate the risk of an FCS Sherman-like dilemma.

Final Thoughts

Given the breadth of cases examined in this study, we can draw a pointed conclusion: Medium-weight forces are useful only when deployed under one or more of the following conditions:

- by air in a way that preempts an effective enemy response (as in Czechoslovakia and Afghanistan)
- against an enemy who lacks the capability to deal with any mobile armor (as in Panama, Somalia, and East Timor)
- in circumstances where other friendly assets—e.g., close air support, artillery, a significant training differential—offset enemy

capabilities (as in Desert Shield and Desert Storm, Angola, and OIF).

In short, this monograph suggests that medium-weight armor enjoys only four clear advantages over heavy armor: rapid deployability (particularly with air-droppable vehicles), speed over roads, trafficability in infrastructure not suited to heavy armor, and lower logistical demands. It furthermore suggests that these advantages are exploitable only in conditions where the resulting diminution of combat power can be accepted or compensated for by other means. Because the U.S. Army cannot expect all future operations to occur in such circumstances, it would be prudent to maintain a mix of heavy, medium-armored, and light forces that can be task-organized and employed in conditions that best match their attributes. Medium-armored forces have much to offer in such a mix.

APPENDIX A

DOTMLPF, BOS, Characteristics of a Transformed Force, and Complex Terrain Synthesis for Case Studies

This appendix synthesizes the DOTMLPF, BOS, characteristics of a transformed force, and complex terrain assessments of the cases examined in this monograph.[1]

DOTMLPF Insights

Table A.1 summarizes the DOTMLPF assessments of all of the cases, showing the particular importance of several factors in the performance—both good and bad—of medium-armored forces: doctrine, organization, training, and materiel. Each case is also discussed individually.

The Centrality of Doctrine

Within the DOMTLPF construct, doctrine plays a central role, as seen in the U.S. Army's own definition of doctrine in FM 3-0, *Operations*:

> Doctrine is the concise expression of how Army forces contribute to unified action in campaigns, major operations, battles, and engagements. While it complements joint doctrine, Army doctrine also describes the Army's approach and contributions to full

[1] The definitions for DOTMLPF, BOS, and characteristics of a transformed force are provided in Appendix C.

Table A.1
DOTMLPF Insights by Case Study

Case	DOTMLPF					
	Doctrine	Organization	Training	Materiel	Leadership and Education	Personnel
Armored warfare in the Spanish Civil War (1936–1939)	X		X	X		
U.S. armored divisions in France and Germany during World War II (1944–1945)	X			X		
Armored cavalry and mechanized infantry in Vietnam (1965–1972)	X	X	X	X	X	
Soviet airborne operations in Prague, Czechoslovakia (1968)	X		X	X		
South Africa in Angola (1975–1988)	X	X	X	X		
Soviet Union in Afghanistan (1979–1989)	X		X	X		X
Operation Just Cause, Panama (1989)		X	X	X		
1st Marine Division light armored infantry in Operation Desert Shield and Operation Desert Storm, Southwest Asia (1990–1991)			X	X		
Russia in Chechnya I (1994–1996)	X	X	X	X	X	X
Russia in Chechnya II (1999–2001)	X	X	X	X		X
Stryker Brigade Combat Teams in Operation Iraqi Freedom (2003–2005)	X	X	X	X	X	
Task Force Ranger in Mogadishu, Somalia (1993)			X	X		
Australia and New Zealand in East Timor (1999–2000)	X	X	X	X		

NOTE: As noted in the methodology section in Chapter 1, this monograph does not assess the "facilities" component of DOTMLPF.

> spectrum operations on land. Army doctrine is authoritative but not prescriptive.[2]

Thus, in the simplest of terms, doctrine is an institutional conceptualization and statement about how things should normally be done, with the expectation that the competent execution of the appropriate doctrine will produce success. Furthermore, future concepts are very often evolutionary, carrying forward the vestiges of existing doctrine that has proven successful (e.g., combined-arms operations) and discarding that which was not successful (e.g., U.S. Army tank destroyers in World War II).

Thus, doctrinal perspectives generally shape decisions in the other domains of DOTMLPF.

Organization

Our analysis of the organizations involved in the historical cases described in Chapters Two through Four yields two principal insights. First, combined-arms organizations that are capable of integrating fire and maneuver, including dismounted infantry, have been critical to conducting successful combat operations, particularly in complex terrain. This was particularly apparent in the hedgerow and urban fighting in World War II and in the Russian experience in Chechnya II. Additionally, the ability to integrate fires outside a service's resources (e.g., the air support provided to U.S. Army forces during World War II and to U.S. Marine Corps LAV units in Operation Desert Storm) can be a critical component of effectiveness, if not survival.

Second, the ability to task organize at relatively low levels to execute specific missions is important. This low-level task organization was apparent in the small, ad hoc combined-arms teams the U.S. Army used in hedgerow fighting in the Normandy *bocage* in World War II and those formed by the Russians during the battle for Grozny in Chechnya II.

---

[2] U.S. Department of the Army, FM 3-0, *Operations*, p. 1-14.

Training

Training generally derives from doctrine. Military institutions develop training regimes and standards to inculcate their forces with the ability to execute doctrine. In the cases examined in this study, the importance of training was evident at two levels. First, well-trained armies executed their doctrine better than poorly-trained ones. Examples of well-trained forces in our case histories include Soviet forces in Czechoslovakia, South African forces in Angola, and U.S. forces in Operation Just Cause, Operation Desert Storm, and OIF. An example of poorly trained forces is provided by the Russian force in Chechnya I, where poorly trained troops failed miserably in the first battle for Grozny.

Second, no amount of training can ensure the success of a force if its doctrine is inappropriate for the situation. This was largely the case with U.S. forces in Vietnam and Task Force Ranger in Mogadishu; it is too soon to tell for the SBCTs in Iraq.

Materiel

Materiel development is heavily influenced by prevailing concepts that become enconded in doctrine. The tanks deployed by the U.S. Army during World War II reflected either the infantry's demand for tanks that were capable infantry support weapons or the cavalry's demand for "iron horses" for traditional cavalry missions. Tank destroyers were created largely in response to General McNair's views about how to deal with the threat posed by the German blitzkrieg. Neither tank nor tank destroyer designs were significantly informed by an appreciation of enemy capabilities. Similarly, the armored vehicles employed in Afghanistan and Chechnya were designed for conventional combat against NATO forces (i.e., for head-to-head combat in direct-fire engagements), where frontal armor is critical, and main guns do not need to engage targets much above or below a horizontal plane. When these vehicles were employed in the mountains or in the urban canyons of Grozny, therefore, many of these vehicles were vulnerable to attack from above, the flanks, and rear, and were incapable of engaging targets.

Finally, constraints—like maximum weight limits—can impede the realization of capability. For instance, a U.S. War Department requirement limited tanks to a weight of 15 tons, preventing tanks from attaining the operational capabilities required by the infantry branch before World War II. A later requirement that U.S. Army tanks not exceed 30 tons further constrained U.S. tank capabilities throughout most of World War II.

In a similar manner, the U.S. Army recently mandated that the interim combat vehicle be transportable by C-130 cargo aircraft. In a decision made partly to help control the vehicle's weight so that this primary requirement could be met, the U.S. Army decided that armor sufficient to protect the vehicle against 14.5-mm machine gun fire would be adequate. In Iraq, where Strykers are now deployed, this level of protection is not sufficient. While slat armor has been added to increase Stryker survivability, it has rendered the vehicle impossible to transport by the C-130.

Quite simply, emerging concepts and desired capabilities—such as rapid deployment by a class of airlift—can place restrictions on developing weapon systems that force critical capability trade-offs. In the case of the Sherman tank and the Stryker vehicle, it was weight constraints that resulted in the construction of vehicles that proved deficient in lethality or survivability in particular combat environments. These outcomes should be considered during the design of the FCS.

Leadership and Education, and Personnel

Leadership and education and personnel were usually not central to the performance of a medium-armored force in the cases examined. Leadership and education are in many ways linked to doctrine because they prepare leaders at various levels to competently plan and execute operations within doctrinal norms.

In the cases examined, personnel was mainly important as a negative: Discipline and quality issues could have a detrimental effect on performance. This was a factor in the Soviet performance in Afghanistan during the latter stages of that conflict, and in Russian performance during Chechnya I. That said, these factors affect performance whether the unit in question is medium-armored, light, or heavy.

BOS Insights

How the BOS operated in the various cases examined in this study was largely a function of decisions made in the constitution of those forces before their initial deployment. In general, the air defense, C2, intelligence, and CSS BOS issues of medium-armored forces did not differ from those of heavy or light forces. However, the case studies do point to distinctions in the maneuver, fire support, and mobility component of mobility/countermobility/survivability BOSs for medium-armored forces when compared to light or heavy forces. BOS insights from the various cases are catalogued in Table A.2.

Maneuver

In several cases, medium-armored forces provided rapid, protected maneuver. In Vietnam, medium-armored forces provided protected maneuver and mobile firepower that reduced casualties from antipersonnel mines and small-arms fire and gave U.S. forces an advantage over VC and NVA light forces. Medium armor provided the Soviet Union with a capability to rapidly topple regimes in Czechoslovakia and Afghanistan. Medium armor afforded coalition forces sufficient maneuverability to rescue the cutoff light infantry of Task Force Ranger in Mogadishu. Finally, the medium-armored forces used by Australia and New Zealand in East Timor allowed their relatively small force to range rapidly through the operational area with protected mobility adequate to the threat condition.

In several cases, however, medium-armored forces were unable to maneuver as expected when they encountered heavier forces or were placed in situations where complex terrain impeded maneuver. German tanks and Italian tankettes were much less lethal against Soviet tanks during the Spanish Civil War, and the Soviet tanks enjoyed a standoff attack advantage. Similarly, U.S. armored forces in World War II were at a disadvantage when confronted with more-capable German armored vehicles, stiff defenses, and capable antitank weapons. In several cases, discussed below, complex terrain contributed to maneuver difficulties.

Table A.2
BOS Insights by Case Study

Case	Battlefield Operating System						
	Maneuver	Fire support	Air defense	Command and control	Intelligence	Mobility/ countermobility/ survivability	CSS
Armored warfare in the Spanish Civil War (1936–1939)	X	X	X	X		X	X
U.S. armored divisions in France and Germany during World War II (1944–1945)	X	X		X		X	
Armored cavalry and mechanized infantry in Vietnam (1965–1972)	X	X			X	X	X
Soviet airborne operations in Prague, Czechoslovakia (1968)	X				X		
South Africa in Angola (1975–1988)	X	X	X	X	X	X	X
Soviet Union in Afghanistan (1979–1989)	X	X				X	X
Operation Just Cause, Panama (1989)	X	X				X	
1st Marine Division light armored infantry in Operation Desert Shield and Operation Desert Storm, Southwest Asia (1990–1991)	X	X					
Russia in Chechnya I (1994–1996)	X	X		X	X	X	X
Russia in Chechnya II (1999–2001)	X	X		X	X		X
Stryker Brigade Combat Teams in Operation Iraqi Freedom (2003–2005)	X	X			X	X	X
Task Force Ranger in Mogadishu, Somalia (1993)	X	X		X		X	
Australia and New Zealand in East Timor (1999–2000)	X						X

During Operation Desert Storm, U.S. Marine Corps medium-armor units recognized the maneuver constraints facing their relatively lightly armored LAV units. Accordingly, they maneuvered LAV units into positions that minimized their vulnerability and allowed them to direct fires against Iraqi units from an advantaged position.

Finally, the maneuverability of the ASLAV-equipped medium-armored forces that the Australian Army deployed to East Timor was constrained; the ASLAVs could not operate off-road in soggy terrain because of the high ground pressure of the wheeled vehicles. The maneuvers of Australian M-113s were not as significantly affected because of their relatively low ground pressure.

Fire Support

In several cases—such as U.S. Army medium-armored forces in northwest Europe and Vietnam and U.S. Marine Corps LAV units in Iraq during Operation Desert Storm—fire support provided medium-armored forces with the crucial edge. Fire support was also a critical BOS for the Russians in both Chechnya cases. It is important to note that fires were often provided by another service (e.g., fixed-wing aviation).

Mobility

Medium-armored forces have contributed to mobility operations in notable ways in two cases. During Operation Just Cause in Panama, M-551 Sheridans were used to breach roadblocks. In Mogadishu, coalition armored vehicles opened up roads to reach cutoff light infantry. In Vietnam, however, an extensive engineering effort was required to clear mines along roads to give U.S. medium forces freedom of maneuver.

Army Transformation Characteristics Insights

Table A.3 shows which cases yielded insights about U.S. Army transformation characteristics in the realms of responsiveness, deployability, agility, versatility, lethality, and survivability. Although we observed sustainabilty characteristics in many cases, sustainabilty issues did not appear to be specific to medium-armored forces. Therefore, they are not addressed separately below.

Table A.3
Army Transformation Characteristics by Case Study

Case	U.S. Army Transformation Characteristics						
	Responsive	Deployable	Agile	Versatile	Lethal	Survivable	Sustainable
Armored warfare in the Spanish Civil War (1936–1939)					X	X	
U.S. armored divisions in France and Germany during World War II (1944–1945)		X	X		X	X	
Armored cavalry and mechanized infantry in Vietnam (1965–1972)			X	X	X	X	
Soviet airborne operations in Prague, Czechoslovakia (1968)		X			X	X	
South Africa in Angola (1975–1988)	X	X			X	X	X
Soviet Union in Afghanistan (1979–1989)		X			X	X	X
Operation Just Cause, Panama (1989)		X			X	X	
1st Marine Division light armored infantry in Operation Desert Shield and Operation Desert Storm, Southwest Asia (1990–1991)			X	X	X		X
Russia in Chechnya I (1994–1996)					X	X	X
Russia in Chechnya II (1999–2001)					X	X	X
Stryker Brigade Combat Teams in Operation Iraqi Freedom (2003–2005)			X	X	X	X	X
Task Force Ranger in Mogadishu, Somalia (1993)					X	X	
Australia and New Zealand in East Timor (1999–2000)	X	X	X	X	X	X	X

Responsive

Medium forces provided a key capability in several conflicts—such as the Soviet coups de main in Czechoslovakia and Afghanistan and the rapid defeat of the Noriega regime in Panama by U.S. light and medium-armored forces—by enabling political leaders to employ military force to rapidly cause the strategic dislocation of the opponent.

Based on its assessment of the threat level in East Timor, Australia decided to employ only medium-armored and light forces in the crisis. This decision allowed Australia to respond much more rapidly than would have been possible if it had deemed heavy forces necessary. Similarly, the SAA's medium-armored forces initially provided South Africa with a very responsive tool in its ongoing involvement in the Angolan Civil War. However, it is difficult to judge the relative utility of medium-armored forces in the above instances because, in each case, the enemy did not fight at all or did not (and could not) fight very well.

In fact, in the case of the Soviets in Afghanistan, the coup de main became a protracted counterinsurgency that the Soviets eventually lost. In the case of the South Africans in Angola, the enemy adapted, causing lethality and survivability issues for South African medium armor; eventually, the SAA fielded an MBT in response.

Deployable

Medium armor capable of airlift or airdrop made critical contributions to Soviet operations in Czechoslovakia and Afghanistan and to U.S. operations in Panama. Deployability was also a critical enabler for South African mechanized forces, who ranged hundreds or thousands of kilometers to the theater of operations, at times by C-130 aircraft. As previously noted, the deployability of Australian and New Zealand medium-armored and light forces enabled those forces to respond rapidly to the crisis in East Timor. However, as was the case with responsiveness, the medium-armored forces in question were able to deploy with such effectiveness because of the operational environment, the relatively modest scale of the conflict, and relative incompetence of the opponents they faced. Since the retirement of the M551 Sheridan, the United States has possessed no air-droppable, forced-entry armor capability.

The case of U.S. medium-armored forces in northwest Europe during World War II provides an example of the negative operational consequences of making deployabilty a principal force design criterion. As previously noted, U.S. Army equipment design was limited by prevailing conceptual views about the employment of armored forces. Shipping and tactical bridging constraints further limited tank and tank destroyer weight limits until late in the war. Consequently, U.S. tanks and tank destroyers operated with an enormous lethality and survivability disadvantage against German tanks and antitank weapons.

Agile and Versatile

Agility and versatility are closely linked. Taken together, they essentially describe the capacity of soldiers, leaders, and organizations to transition between changing situations and operate with equal effectiveness across the spectrum of operations. Clearly, agility and versatility affect the entire DOTMLPF spectrum because they imply a great deal of organic capacity for adaptation and flexibility.

As previously discussed, conceptual decisions about the purpose of medium-armored forces can limit their agility and flexibility. U.S. medium-armored forces in Vietnam were fundamentally created to fight conventional wars. Although there was considerable adaptation of TTP to local conditions throughout the war, it is arguable that none of these adaptations prepared units to conduct the counterinsurgency operations that were an important dimension of the war. On the other hand, Australian and New Zealand forces adapted well to the conditions in East Timor and were effective in their peacemaking role, which was not what they were designed to do. One clear difference between the two cases was the much higher level of violence in Vietnam than in East Timor. A second difference is that U.S. forces were fighting both an insurgency against the VC and a conventional war against competent and well-armed NVA forces. That said, conventional-unit operations in Vietnam were generally focused on closing with and destroying the enemy—VC or NVA regular—in a war of attrition.

Some evidence suggests that, at the beginning of postwar operations in OIF, the approach used by U.S. forces, including the SBCTs, was focused more on offensive operations to find and neutralize insurgents than on protecting and winning the support of the Iraqi people. This approach has changed over time and the quality of the forces in the SBCTs seems to give the teams a great deal of agility and versatility.[3]

Lethal and Survivable

Lethality and survivability are closely related in the design of armored fighting vehicles. Lethality is the ability to kill or neutralize adversaries. Long-range target acquisition, accurate fire control, and effective stand-off weapons all contribute to lethality. Many of the capabilities that contribute to lethality also aid survivability. Therefore, the U.S. Army emphasizes "the capability to see first, understand first, act first, and finish decisively."[4] Thus, survivability is directly tied to being able to kill or neutralize an opponent before being killed or neutralized yourself.

The starkest example of the linkage between lethality and survivability vis-à-vis an opponent was the case of U.S. versus German armored forces in World War II. The combination of more-powerful main guns and better frontal armor gave the Germans a decided advantage over U.S. armored forces. U.S. medium tanks and tank destroyers were vulnerable at much longer ranges than first-line German armored vehicles. Mass, coupled with firepower from artillery and airpower, was the great lethality equalizer.

In conflicts that are not conventional, however, survivability is also affected by the ability of an adversary to attack armored vehicles where they are weakest—their sides, top, and rear. In the majority of the cases assessed in this study, all armored vehicles (both medium-armored and heavy) were vulnerable to close-range attack by man-

---

[3] Thomas E. Ricks, *Fiasco: The American Military Adventure in Iraq*, New York: The Penguin Press, 2006. Ricks describes the mixed performance of several units in dealing with the postwar situation in Iraq and the costs. He also discusses the efforts of the U.S. Army to learn and adapt.

[4] U.S. Department of the Army, *Army Transformation Wargame 2001*, Fort Monroe, Headquarters, U.S. Army Training and Doctrine Command, n.d., p. 3.

portable antitank weapons and to mines. Enemies wielding these weapons often denied armored forces the ability to "see first," particularly in complex terrain. In-the-field modifications, ranging from welding on extra armor plate to providing slat armor, were often applied to counter these threats.

Complex Terrain Insights

Table A.4 shows the various types of complex terrain encountered by the medium-armored forces examined in this study. It is important to note that in almost every case examined in this monograph, medium-armored forces had to operate in some form of "close" complex terrain. Furthermore, most of the operations were affected by underdeveloped infrastructure. These factors generally place higher demands on CSS functions because of heavier maintenance demands on equipment and the absence of host nation support. And, in the case of several protracted conflicts (Afghanistan, Chechnya, OIF), the infrastructure deteriorated further as the war continued.

Naturally, complex terrain frequently restrained the tempo of maneuver. There is, however, another important dimension of complex terrain that is evident in many of the cases: Complex terrain imparts an advantage to the defender and often creates survivability problems for the attacker.

This is true for two reasons. First, most armored vehicles, including medium-armored vehicles, are designed to face other armored vehicles head-on and be most strenuously attacked from the front where their armor is therefore thickest. In complex terrain, however, direct-fire attacks frequently occur at close range and are aimed at the more vulnerable sides, rear, or tops of vehicles. Additionally, vehicles are vulnerable to mines and IEDs, weapons that are easier to conceal in complex terrain. Furthermore, complex terrain marginalizes the lethality advantage usually afforded by standoff fires and degrades the ability to see first.

Second, in the urban canyons that characterize many large cities, the ability to employ weapons from protected firing positions (and

Table A.4
Complex Terrain Experience by Case Study

Case	Complex Terrain: Urban	Mountainous	Jungle	Forests	Hedgerows	Undeveloped infrastructure
Armored warfare in the Spanish Civil War (1936–1939)	X	X				
U.S. armored divisions in France and Germany during World War II (1944–1945)	X	X		X	X	
Armored cavalry and mechanized infantry in Vietnam (1965–1972)		X	X			X
Soviet airborne operations in Prague, Czechoslovakia (1968)	X					
South Africa in Angola (1975–1988)				X		X
Soviet Union in Afghanistan (1979–1989)		X				X
Operation Just Cause, Panama (1989)	X					
1st Marine Division light armored infantry in Operation Desert Shield and Operation Desert Storm, Southwest Asia (1990–1991)						X
Russia in Chechnya I (1994–1996)	X	X				X
Russia in Chechnya II (1999–2001)	X	X				X
Stryker Brigade Combat Teams in Operation Iraqi Freedom (2003-2005)	X					X
Task Force Ranger in Mogadishu, Somalia (1993)	X					X
Australia and New Zealand in East Timor (1999–2000)			X			X

at very high angles of elevation) is important. This was particularly apparent in Grozny, where Russian forces eventually had to deploy air-defense guns to be able to engage Chechen fighters in the upper stories of buildings. It was also an important consideration in the mountainous terrain that characterized much of the Soviet fighting in Afghanistan.

Finally, operations in urban and mountainous terrain require weapons of medium caliber (20 to 35 mm) that can be fired on the move, offer high rates of accurate fire, and accommodate explosive ammunition.

APPENDIX B

Individual Case Study Assessments of DOTMLPF, BOS, Characteristics of a Transformed Force, and Complex Terrain

This appendix contains detailed assessments of the DOTMLPF, BOS, characteristics of a transformed force, and complex terrain insights gleaned from each historical case analyzed in the main body of the monograph. Although it repeats some material from earlier chapters, this appendix presents the analytical insights in a distilled format and offers supplementary historical information.

Medium-Armored Forces in Large-Scale Combat Operations

Armored Warfare in the Spanish Civil War (1936–1939)

DOTMLPF Insights. This case offers insights in the areas of doctrine, training, and materiel.

Doctrine. Each nation that provided armored forces to the Spanish Republicans and Nationalists produced views about how to employ large, massed, armored formations in rapid, decisive operations. The Germans were developing the blitzkrieg.[1] The Soviets had embraced concepts stressing "decisive victory . . . by offensive action in depth" that were codified in the Provisional Field Service Regulations of 1936.[2]

[1] Daley, "The Theory and Practice of Armored Warfare in Spain," pp. 39–40.

[2] Daley, "The Theory and Practice of Armored Warfare in Spain," p. 40.

The Italians became committed to a theory of *guerra di rapido corso*.[3] The stabilized conditions that existed when foreign formations intervened in the war, coupled with the relatively small numbers of armored vehicles deployed, created circumstances where these various theories of rapid, decisive operations could not be executed. Instead, armored vehicles became tactical weapons normally employed in support of limited offensive operations or to bolster defenses.

What did develop over time in Spain was an appreciation among all the forces involved of the importance of the contributions of various arms in offensive combat. This was reported by a U.S. officer in 1939:

> In most offensive operations carried out by either side during the past year or more, tanks seem to have been the third echelon of the attack. Aviation and artillery strike the first and second blows, tanks the third, infantry the fourth, and cavalry enters the action as the fifth and final echelon to pursue, outflank, or mop-up.[4]

The same author noted, however, that tanks had shown some value in pursuit and as a counterattack force "if used before the enemy has organized the newly won terrain and brought forward his antitank weapons."[5] Nevertheless, one lesson was very clear: Armored forces, even during limited breakthroughs and exploitations achieved during the Spanish Civil War, required competent infantry support to negate antitank defenses.[6] Thus, combat in Spain showed that

> whatever promise independent mechanized action held at the operational and strategic levels, frequent combined-arms operations involving tanks and dismounted infantry were to be expected.[7]

---

[3] Sullivan, "The Italian Armed Forces," in Millet and Murray, eds., *The Interwar Period*, p. 706.

[4] Johnson, "Employment of Supporting Arms," p. 13.

[5] Johnson, "Employment of Supporting Arms," p. 13.

[6] Daley, "Soviet and German Advisors Put Doctrine to the Test," p. 36.

[7] Daley, "The Theory and Practice of Armored Warfare in Spain," p. 42. See also Candil, "Soviet Armor in Spain," p. 38. The author concluded in this article that the Spanish Civil

Antony Beevor believes that the Germans learned something at the higher level of operations:

> Their tanks needed to be more heavily armed and concentrated in armoured divisions for 'Schwerpunkt' breakthroughs. . . . [I]t was as a result of the war in Spain that the German army saw the need to increase the size and power of its tank force.[8]

The effect of the war on Soviet concepts for armored warfare was much more constrained:

> The purging of Marshal [Mikhail Nikolayevich] Tukhachevsky and his supporters who advocated the new approach to armored warfare returned communist military theory to the political safety of obsolete tactics.[9]

Thus, even during the war in Spain, "Soviet advisers could not advocate modern armoured tactics after the show trial of Marshal Tukhachevsky."[10]

Training. Training problems arose in all the armored forces during the Spanish Civil War, particularly in the German *Imker Drohne* and in the Soviet Krivoshein Detachment. Both of these units had to rely on Spanish volunteers for tank crewmen, and according to German commander Wilhelm von Thoma, these volunteers "were 'quick to learn' but also 'quick to forget' how to operate tanks."[11] The language barrier proved formidable for the Germans and Soviets, but particularly so in the Krivoshein Detachment—none of the Soviet instructors spoke

---

War experience showed that "[t]anks needed to be supported by motorized infantry. Failing to do that caused many of the Soviet mistakes. Only in rare cases, or against limited objectives, should tanks proceed alone." Furthermore, "[a] great advantage accrued to close cooperation with aircraft, which could air command and control, provide combat support, and perform reconnaissance."

[8] Beevor, *The Battle for Spain*, p. 427.

[9] Beevor, *The Battle for Spain*, p. 196.

[10] Beevor, *The Battle for Spain*, p. 427.

[11] Daley, "The Theory and Practice of Armored Warfare in Spain," p. 42.

Spanish and crews had to be trained through interpreters.[12] Further complicating the Krivoshein Detachment's training difficulties was the issue of recruiting politically reliable Spaniards for training:

> Because the Soviet T-26 was a concrete manifestation of proletarian revolutionary might, only devout Communists were allowed to operate it. . . . [N]on-Communists with mechanical backgrounds were often rejected in favor of more politically acceptable but technically unqualified inductees.[13]

Finally, deficient collective training within units and among the various arms operators hampered operations.[14]

Materiel. The Spanish Civil War offered several lessons for the design of armored vehicles. The lightly armored tanks employed in the conflict were vulnerable to other tanks and to antitank weapons, ranging from antitank guns to field expedient devices (such as what would become known as Molotov cocktails). The conditions in Spain of "battle-torn terrain" and "natural and artificial obstacles" made the going very difficult for tanks and made speed "unusable."[15] Thus, the need for protection militated for armor over speed: "Most foreign commentators now stress armor above speed. Certainly, if one or the other has to be sacrificed, speed must give way to armor."[16] Additionally, it became apparent that tanks needed turrets that could traverse 360 degrees (to address flank threats) and guns and accurate fire control systems capable of dealing with other tanks. On-board tank radios were necessary to command and control tank units.[17] Finally, the superiority of the Soviet T-26 and BT-5 in the Spanish Civil War derived from the greater lethality at longer ranges of their main guns when compared to

---

[12] Daley, "The Theory and Practice of Armored Warfare in Spain," pp. 42–43.

[13] Daley, "The Theory and Practice of Armored Warfare in Spain," p. 43.

[14] Daley, "Soviet and German Advisors Put Doctrine to the Test," p. 36.

[15] Johnson, "Employment of Supporting Arms," pp. 13–14.

[16] Johnson, "Employment of Supporting Arms," p. 16.

[17] Daley, "The Theory and Practice of Armored Warfare in Spain," pp. 41–42.

the German PzKpfw I and Italian tankettes, a lesson that would also affect tank design in the coming Second World War.

BOS Insights. This case yields BOS insights in the areas of maneuver, fire support, air defense, C2, mobility and survivability, and CSS.

Maneuver. As previously noted, the Spanish Civil War demonstrated the importance of coordinated armor and infantry formations. When tanks became separated from infantry, they became vulnerable to antitank measures.

Fire Support. Fire support became increasingly sophisticated as the war continued. Aviation, artillery, and tank fires were employed to support offensives. Furthermore, air attack became a serious threat to ground forces.[18] Artillery, in particular, proved its worth:

> Despite the employment of tanks and the direct participation of combat aviation in the ground battle, it has been proved in this war that without an ample supply of efficiently handled artillery, well supplied with munitions, there is no advance. Men cannot face un-neutralized machine-gun fire, nor weak tanks oppose unsilenced antitank guns.[19]

Air Defense. The evolving threat from the air resulted in passive and active air-defense measures. Ground forces began to practice "dispersion, concealment, camouflage and cover."[20] Pursuit-aircraft operations and the quality of air-defense weapons improved markedly over their World War I counterparts. Airplanes were thus forced to operate at higher altitudes.[21] By the end of the war, air-defense quality was "measured by the degree to which it prevents or defeats hostile aerial activity directed at ground elements rather than by its destruction o[f] aircraft."[22]

---

[18] Johnson, "The Employment of Supporting Arms," p. 12.

[19] Johnson, "The Employment of Supporting Arms," p. 12.

[20] Johnson, "The Employment of Supporting Arms," p. 20.

[21] Johnson, "The Employment of Supporting Arms," p. 20.

[22] Johnson, "The Employment of Supporting Arms," p. 20.

Command and Control. As previously noted, C2 were severely taxed by the increased mobility imparted to the battlefield by mechanized vehicles and the need to synchronize fire and maneuver. The fact that few tanks or airplanes were equipped with radios during the conflict exacerbated this problem. C2 were further complicated by uneven levels of training.[23]

Mobility/Countermobility/Survivability. Combat engineering was an important component of operations during the Spanish Civil War. In the area of mobility, engineers "had their hands full keeping roads open, installing ponton [*sic*] bridges, and repairing permanent bridges."[24] Survivability tasks included the construction of field works, antimechanized defenses, obstacles, and minefields.[25]

Combat Service Support. Logistics constrained and hampered operations throughout the Spanish Civil War. The tanks employed in the conflict, for example, were not much advanced over their World War I predecessors, and proved mechanically unreliable. A contemporary author noted that tanks

> must be built as mechanically perfect, yet as simple, as technical science can make them; they must be regularly serviced and overhauled; must be handled by experts; and must be conserved for their most effective uses.[26]

Inadequate tank maintenance and recovery systems, combined with the inherent unreliability of the tanks used in Spain, resulted in the abandonment of many of the tanks that broke down during

---

[23] Johnson, "The Employment of Supporting Arms," p. 9. Radios contributed to what are now known as psychological operations. Johnson notes that

> [o]ne major use of radio is for propaganda purposes. Both sides have powerful stations from which come continual broadcasts for local and foreign listeners. Loud speakers—the *altavoz*—blare across no-mans-land telling the hostile combatants of the villainy of their superiors and the virtues of their opponents, of the utopia awaiting deserters and the purgatory in store for the rest.

[24] Johnson, "The Employment of Supporting Arms," p. 9.

[25] Johnson, "The Employment of Supporting Arms," pp. 9, 14–15.

[26] Johnson, "The Employment of Supporting Arms," p. 16.

operations. It appears that this was a more significant problem for the Republicans than for other combatants. By 1939, few of the Soviet tanks sent to Spain remained operational.[27]

Characteristics of Transformation Insights. This case yields transformation insights in the areas of lethality, survivability, and sustainability.

Lethality. The war clearly proved that tanks armed with cannons, like the Soviet T-26 with its 45-mm gun, were capable of destroying other armored vehicles at ranges that made them invulnerable to the machine guns used by their opponents.

Survivability. They very light armor of all of the armored vehicles used in the Spanish Civil war was vulnerable to a wide range of threats. One observer noted the following:

> Tanks have not become impotent through the capabilities of defensive means against them. But it should be realized that tanks now have to fight on an equal basis with antitank defense. Obstacles, traps, mines, antitank guns, and the enemy's own armored fighting vehicles, these are to tanks what trenches, barbed wire, artillery and machine-gun fire are to foot or mounted troops.[28]

To survive antitank measures, tanks used air, artillery, or infantry to destroy or neutralize the threats. Against other tanks, the story of tank survivability was mixed. Soviet tanks could use the stand-off range afforded by their 45-mm cannons to attack German Pzkpfw I and Italian tankettes. Thus, cannon range gave Soviet tanks a survivability advantage. Although they proved useful as infantry support guns, German and Italian tanks "stood no chance in tank-versus-tank combat against Republican opponents."[29] The tanks and tankettes survived by avoiding Soviet tanks and armored cars.

---

[27] Johnson, "The Employment of Supporting Arms," p. 16; Daley, "Soviet and German Advisors Put Doctrine to the Test," pp. 35–36.

[28] Johnson, "The Employment of Supporting Arms," p. 17.

[29] Daley, "The Theory and Practice of Armored Warfare in Spain," p. 42.

Complex Terrain Insights. Tanks were used throughout the theater of operations during the war. Fighting occurred in the mountains, in cities, and on broken terrain; mechanized operations were complicated by deliberately planted obstacles. These conditions highlighted the importance of coordinated combined-arms operations, particularly between tanks and infantry. Furthermore, the relatively crude machines used by the Nationalists and the Republicans were subject to breakdown, were easily "ditched," and thus were frequently roadbound.

Tanks also participated in operations in villages and cities. On the offensive, given their light armor, they were "most vulnerable to grenades and often makeshift antitank measures."[30] Tanks were useful, however, in a "fire brigade" role in cities. This was particularly apparent in the extended Republican defense of Madrid, where Soviet T-26s "were mobile enough to appear at any threatened point and well enough armed to make a crucial difference once there."[31]

U.S. Armored Forces versus German Armored Forces in Western Europe During World War II (1944–1945)

DOTMLPF Insights. This case offers insights about U.S. Army medium armor in the areas of doctrine, organization, and materiel. We first explain the pre-1944 origins of these DOTMLPF elements to provide important context for the insights that follow.

The U.S. Army's experience with armored forces began in World War I. During the Great War, the U.S. Army created a Tank Corps and fielded units that saw combat. The 304th Tank Brigade, under the command of Lieutenant Colonel George S. Patton, Jr., participated in the September 1918 offensive to reduce the St. Mihiel salient and in the final Allied Meuse-Argonne offensive that began on September 26, 1918, and ended with the Armistice in November. The U.S. components of the 304th consisted of the 344th and 345th Light Tank Battalions, which were equipped with French-supplied 7.4-ton Renault tanks. The other U.S. unit, the 301st Heavy Tank Battalion, was assigned to the

[30] Daley, "Soviet and German Advisors Put Doctrine to the Test," p. 34.

[31] Daley, "Soviet and German Advisors Put Doctrine to the Test," p. 36.

British 2nd Tank Brigade. This battalion, equipped with British Mark V heavy tanks and assigned to the British 2nd Brigade, supported the II American Corps in its assault on the Hindenburg line in late September 1918.[32]

The wartime experience of the U.S. Army Tank Corps was mixed. World War I tanks were slow, mechanically unreliable, and traversed the shell-pocked battlefields with difficulty; many broke down or became "ditched" in action. The 304th Tank Brigade began the Meuse-Argonne offensive with 142 Renault tanks. By November 1, 1918, it could field only 16 tanks to support the final assault.[33]

In the aftermath of World War I, U.S. Army reorganization legislation abolished the Tank Corps, largely because U.S. Army senior leadership, most notably General John J. Pershing, viewed the tank as an infantry support weapon. Leadership therefore believed that future development of the tank should be left to infantry. Under the provisions of the National Defense Act of 1920, the newly constituted infantry branch was given responsibility for tanks, including the promulgation of doctrine and the establishment of materiel requirements. In 1931 the cavalry branch received War Department authority to develop tanks, which were known as "combat cars" to avoid the strictures of the National Defense Act.

Three critical factors affected the development of U.S. tanks: branch parochialism, weight constraints, and competing ideas about how to defeat enemy armor. Branch parochialism resulted in two categories of tanks: those designed as infantry support weapons, and those used as "iron horses" focused on traditional cavalry missions. Weight constraints affected what could be accomplished in tank design within the competing demands of speed, lethality, and protection. An increase in any of the three caused an increase in weight. A U.S. Army–imposed weight ceiling required trade-offs in one of the other three. Initially, the U.S. Army chief of engineers set the maximum weight for tanks at 15 tons, selecting this limit to match the carrying capacity of the

[32] Johnson, *Fast Tanks and Heavy Bombers*, pp. 36–37.

[33] Johnson, *Fast Tanks and Heavy Bombers*, p. 35. See also Wilson, *Treat 'Em Rough!*, for a discussion of the World War I U.S. Army Tank Corps and its demise.

divisional pontoon bridge.[34] Throughout the interwar period the U.S. Army Ordnance Department struggled unsuccessfully to provide the infantry and cavalry with tanks and combat cars that met their requirements within this weight limitation.[35] The U.S. Army increased tank weight limits later on, but still set a maximum (of 30 tons, with a width of 103 inches) to facilitate shipping and to ensure "that navy transporters and portable bridges did not need to be redesigned in the midst of the war."[36]

The question of how to use armored forces was also an issue in the U.S. Army. The chief of infantry believed that tanks existed to support attacking infantry. The 7th Cavalry Brigade (Mechanized), stationed at Fort Knox, Kentucky, willingly embraced armor, viewing combat cars as "iron horses." This brigade believed that tanks should be used, like traditional cavalry, to exploit and pursue infantry breakthrough attacks to complete the defeat of an enemy in depth. The chief of the Army Ground Forces, Lieutenant General Lesley J. McNair, was convinced that the appropriate response to the massed armored attacks employed in the German blitzkrieg offensives at the beginning of World War II was the tank destroyer.

Not until June 1940, after the success of the German blitzkrieg in Poland and France, did the U.S. Army merge its existing infantry and cavalry armored units into an Armored Force. The first chief of the Armored Force was Brigadier General Adna R. Chaffee, Jr., commander of the 7th Cavalry Brigade (Mechanized). At the critical period of the initial formation of the Armored Force and the armored division, General Chaffee put a distinctive cavalry stamp on the Armored Force.[37]

---

[34] Johnson, *Fast Tanks and Heavy Bombers*, p. 74.

[35] Johnson, *Fast Tanks and Heavy Bombers*, pp. 80, 200–201.

[36] House, *Combined Arms Warfare in the Twentieth Century*, p. 152. See also Baily, *Faint Praise*, p. 127. The 30-ton weight requirement in AR 850-15 was not relaxed until late 1944, when it was waived for the fielding of the T-26 tank. Because of its weight, the T-26 could not cross U.S. Army tactical bridges then in the field.

[37] House, *Combined Arms Warfare in the Twentieth Century*, p. 152.

During World War II, the U.S. Army fielded 16 armored divisions in Europe. Additionally, the U.S. Army put 65 independent tank battalions into the field to support its infantry divisions, compared to the 54 tank battalions within its armored divisions.[38] The U.S. Army had also fielded 61 tank destroyer battalions in Italy and western Europe by war's end.[39] These forces were equipped with the weapons shown in Table 2.2.

Doctrine. Doctrine for U.S. Army tanks and tank destroyers evolved from the creation of the Armored Force in 1940 and the Tank Destroyer Center in 1941.[40] In both cases, steady institutional evolution as well as bottom-up innovation by units in the field throughout the war contributed to the development of doctrine.

Tank doctrine. There were in essence two tank doctrines in the U.S. Army, each reflecting the interwar influence of infantry and cavalry perspectives on how best to employ armor. Thus, the U.S. Army fielded tank units to support the infantry divisions and serve as the principal units in armored divisions.

The nondivisional independent tank battalions were organized to support the mission that was aiding the advance of the infantry. The 1940 edition of FM 100-5, *Operations*, specified that these battalions would attack in two echelons. The first echelon sought to destroy enemy antitank guns; the second provided support to attacking infantry.[41] Cooperation between these tank battalions and supported infantry divisions continually improved because of their habitual association. It became standard practice to assign a tank battalion to infantry divisions and, generally, the companies of the battalion were spilt up to support the division's infantry regiments. Thus, "[f]or all practical

---

[38] Gabel, "World War II Armor Operations in Europe," in Hofmann and Starry, eds., *From Camp Colt to Desert Storm*, p. 155.

[39] Gabel, "World War II Armor Operations in Europe," in Hofmann and Starry, eds., *From Camp Colt to Desert Storm*, p. 178.

[40] Baily, *Faint Praise*, p. 16.

[41] Johnson, *Fast Tanks and Heavy Bombers*, pp. 145–46.

purposes the tank company became an organic part of the infantry regiment."[42]

The formative doctrine of the early armored divisions derived from vintage cavalry doctrine and focused on "dash and speed rather than combined arms."[43] The March 1942 *Armored Force Field Manual* specified that the role was

> the conduct of highly mobile ground warfare, primarily offensive in character, by self-sustaining units of great power and mobility, composed of specially equipped troops of the required arms and services.[44]

Consequently, "[a]rtillery and infantry were subordinated to supporting roles—fixing the enemy and occupying captured positions."[45] The manual further "emphasized surprise, speed, shock action, and firepower directed against rear areas," and its "preferred tactics for armored formations were breakthrough, exploitation, encirclement, annihilation, and pursuit."[46] This doctrine

> was predicated on the assumption that tanks operated in masses, at their own pace, and that combined arms consisted of attaching supporting, subordinate elements to armored regiments.[47]

---

[42] Gabel, "World War II Armor Operations in Europe," in Hofmann and Starry, eds., *From Camp Colt to Desert Storm*, pp. 162–163.

[43] Gabel, "World War II Armor Operations in Europe," in Hofmann and Starry, eds., *From Camp Colt to Desert Storm*, p. 146.

[44] Gabel, "World War II Armor Operations in Europe," in Hofmann and Starry, eds., *From Camp Colt to Desert Storm*, p. 147.

[45] Gabel, "World War II Armor Operations in Europe," in Hofmann and Starry, eds., *From Camp Colt to Desert Storm*, p. 149.

[46] Gabel, "World War II Armor Operations in Europe," in Hofmann and Starry, eds., *From Camp Colt to Desert Storm*, p. 147.

[47] Gabel, "World War II Armor Operations in Europe," in Hofmann and Starry, eds., *From Camp Colt to Desert Storm*, p. 143.

As the war progressed, the U.S. Army applied lessons learned from the field, particularly North Africa, and the doctrine for U.S. armored divisions took on more of a combined-arms tone. Thus, the January 1944 version of FM 17-100, *Armored Command Field Manual: The Armored Division*, "stressed the need for timely cooperation among the arms while placing more emphasis on the destruction of enemy forces in contact and less on cavalry-like rampages in hostile rear areas."[48] But combined arms in this manual referred to U.S. Army Ground Force units and did not envision the integration of air power to the levels realized in the German blitzkrieg. Furthermore, even if envisioned, air-ground cooperation would have been problematic, given the reality that the U.S. Army Air Forces were focused on strategic bombing and, as an institution, were not keen on the idea of subordinating air power to ground forces. Instead, ad hoc procedures developed in the combat theaters to provide air support to ground units.[49]

Finally, one glaring deficiency in U.S. armored doctrine, both in the independent tank battalions and the armored divisions, was the assumption that tanks would not fight enemy tanks:

> The main purpose of the tank cannon is to permit the tank to overcome enemy resistance and reach vital rear areas, where the tank machine guns may be used most advantageously.[50]

Therefore, although there was recognition that "[c]hance encounters between tanks would occur . . . the principal role of the armored division was to exploit and pursue, not to fight enemy armor."[51] If required, "antimechanized protection" would be provided by attaching tank destroyer units.[52]

---

48 Gabel, "World War II Armor Operations in Europe," in Hofmann and Starry, eds., *From Camp Colt to Desert Storm*, p. 147.

49 Johnson, *Fast Tanks and Heavy Bombers*, p. 226.

50 Johnson, *Fast Tanks and Heavy Bombers*, p. 226.

51 House, *Combined Arms Warfare*, p. 152.

52 U.S. War Department, FM 17-100, *(Tentative) Employment of the Armored Division and Separate Armored Units*, pp. 1, 8, 13.

Unfortunately for U.S. tankers, and in spite of U.S. Army doctrine, U.S. tanks did have to fight German tanks and did so at a great disadvantage. Most tank engagements were small actions. Historian Charles Baily notes that the 2nd Armored Division's biggest tank battle through the end of World War II "involved only twenty-five German tanks."[53] This action occurred in mid-November 1944 in the vicinity of Puffendorf, Germany. Over a two-day period, the U.S. 1st Battalion, 67th Armored Regiment, 2nd Armored Division, suffered 363 casualties and lost 57 tanks to well-sited German tanks. The battalion claimed only four German tanks destroyed—two by Shermans and two by M36 tank destroyers.[54]

Tank destroyer doctrine. U.S. Army doctrine envisioned two principal roles for tank destroyers. First, they supported offensive operations by protecting friendly forces from enemy armored counterattacks. Second, they supported defensive operations by defending in depth against enemy armor attacks, with the majority of tank destroyers retained in a mobile reserve to respond to the main enemy attack.[55] This latter role was the main U.S. response to the German blitzkrieg.

In only one instance during World War II did a U.S. tank destroyer battalion ever execute its prescribed doctrine. During a March 1943 engagement near El Guettar in North Africa, the 601st Tank Destroyer Battalion (composed of M3 tank destroyers), with an attached company from the 899th Tank Destroyer Battalion (composed of M10 tank destroyers), turned back a German force of some 50 Panzers, but with heavy losses: 20 of 28 M3s and seven of ten M10s were destroyed.[56]

In northwest Europe, U.S. forces rarely encountered large German armored formations. Instead, the norm was the tough business of

[53] Baily, *Faint Praise*, p. 92.

[54] Johnson, *Fast Tanks and Heavy Bombers*, p. 195.

[55] Johnson, *Fast Tanks and Heavy Bombers*, p. 150.

[56] Gabel, "World War II Armor Operations in Europe," in Hofmann and Starry, eds., *From Camp Colt to Desert Storm*, p. 152.

"[p]rying German infantry and guns from well-prepared positions."[57] In practice, much like the independent tank battalions, tank destroyer battalions were semipermanently assigned to U.S. divisions and their companies were task-organized with infantry regiments.[58]

Organization. The organization of the U.S. armored division evolved throughout the war and experienced six different changes, two of which were "most significant."[59] The greatest differences between the 1940 organization and later armored divisions (the 1942 "heavy" and the 1943 "light") were (1) the types of command echelon between the division and its maneuver elements and (2) the steadily increasing balance between tanks and infantry within the division.[60] The 1940 division contained an armored brigade, which had two light-armored regiments, one medium-armored regiment, and a field-artillery regiment. Its infantry was organized in a separate regiment containing two battalions.[61] The division's structure reflected a bias towards cavalry-type missions, in that it had "287 light tanks and 120 mediums organized into six light battalions and two medium battalions . . . [and] only two battalions of infantry."[62] The high preponderance of light tanks in the division reflected a U.S. Army emphasis on mobility over protection or firepower, while the relatively low infantry strength demonstrated both the division's supporting role and a lack of emphasis on combined arms.[63]

---

57 Baily, *Faint Praise*, p. 114.

58 Gabel, "World War II Armor Operations in Europe," in Hofmann and Starry, eds., *From Camp Colt to Desert Storm*, p. 163.

59 House, *Combined Arms Warfare*, p. 139. Throughout the war, all armored divisions included engineer, field artillery, and support units.

60 House, *Combined Arms Warfare*, pp. 139–141.

61 John B. Wilson, *Maneuver and Firepower: The Evolution of Divisions and Separate Brigades*, Washington, D.C.: Center of Military History, 1998, p. 151. This action occurred in March 1943 near El Guettar in North Africa.

62 Gabel, "World War II Armor Operations in Europe," in Hofmann and Starry, eds., *From Camp Colt to Desert Storm*, p. 146.

63 Gabel, "World War II Armor Operations in Europe," in Hofmann and Starry, eds., *From Camp Colt to Desert Storm*, pp. 147–149; see also p. 167, which notes that the light tanks

A 1942 revision to the armored division removed the armored brigade from the 1940 organization and substituted

> two Combat Commands, A and B (CCA and CCB), headquarters that might control any mixture of subordinate battalions given them for a particular mission. . . . The 1942 organization also reversed the ratio of medium and light tanks, leaving the armored division with two armored regiments, each consisting of one light and two medium tank battalions. The new structure still had six tank battalions but only three armored infantry and three armored field artillery battalions.[64]

In 1943 the U.S. Army adopted a new, smaller armored-division organization. It removed regiments from the division and had

> three battalions each of tanks, armored infantry, and armored field artillery, although in practice there were still twelve tank companies to only nine infantry. A third, smaller combat headquarters, designated reserve (R), was added to control units under division control and not currently subordinated to the other two combat commands. Some division commanders used this CCR as a third tactical control element like CCA and CCB.[65]

The U.S. Army fielded 16 armored divisions in World War II. Two of these divisions, the 2nd and 3rd, retained the 1942 organizational structure, and were called "heavy armored divisions"; the remaining 14 divisions adopted the 1943 "light armored division" organization.[66] The incorporation of combat commands within the armored division greatly facilitated combined arms task organization of the divisions' organic units and the attachment of non-divisional units.

---

within the armored divisions were generally relegated to reconnaissance, screening, and other roles because of their vulnerability.

[64] House, *Combined Arms Warfare*, pp. 140–141.

[65] House, *Combined Arms Warfare*, p. 141.

[66] Gabel, "World War II Armor Operations in Europe," in Hofmann and Starry, eds., *From Camp Colt to Desert Storm*, p. 155.

Training. The U.S. Army was a mobilized force that grew from a 1940 strength of 269,023 officers and men to a peak of 8,227,958 in 1945.[67] Deep experience and high levels of training could not be expected in such an army. The 12th Army Group postwar after-action review noted the following deficiencies in combined-arms training:

> Our infantry-tank combined training in the US was inadequate. Close knitting of this team is mandatory. Adequate training requires that officers of all grades in the infantry and armored units are thoroughly cognizant of the capabilities and limitations of each arm. This cannot be satisfactorily conducted through the medium of textbooks or academic courses. The only satisfactory training is through indoctrination over a long period by means of combined combat exercises.[68]

Another clear training deficiency was evident in the area of air-ground operations. This deficiency arose partly from the fact that ground doctrine did not rely heavily on air support. Another reason was the reluctance of the U.S. Army Air Forces to divert resources away from their institutional priorities, most notably the strategic bombing campaign. They were therefore "unwilling to provide aircraft even for major ground maneuvers, let alone for small-unit training."[69] Consequently, training in air-ground operations suffered, and

> [s]ix months before the Normandy invasion . . . thirty-three U.S. divisions in England had experienced no joint air-ground training. . . . [A]ir and ground units went overseas with little understanding of the tactics and capabilities of their counterparts.[70]

Materiel. U.S. Army tank and tank destroyer design followed doctrinal and institutional preferences. The light tanks envisioned as

67 Russell F. Weigley, *History of the United States Army*, enlarged ed., Bloomington, Ind.: Indiana University Press, 1984, p. 599.

68 U.S. 12th Army Group, *12th Army Group Report of Operations*, pp. 59–60.

69 House, *Combined Arms Warfare*, p. 170.

70 House, *Combined Arms Warfare*, p. 170.

"iron horses" in early tank-division designs were relegated to a reconnaissance role by the time of the Normandy invasion. U.S. medium-armored tanks and tank destroyers, however, were vastly outclassed by many German vehicles (see Table 2.3). More on this issue is provided in the lethality and survivability discussions, below.

BOS Insights. This case yields BOS insights in the areas of maneuver, fire support, and C2.

Maneuver. The types of maneuver anticipated by U.S. doctrine rarely occurred in northwest Europe. The most notable exception was the breakout from the Normandy beachhead and its exploitation during Operation Cobra in July 1944. Allied carpet bombing created a rupture in the German lines and U.S. ground forces broke out. Over the next several weeks the Allies swept through northern France and Belgium, and the "six American armored divisions in theater during the pursuit were in their glory."[71] The exploitation ground to a halt in early September:

> Logistical overextension, geography, weather, and a resurgent German defense all combined to stop the dash toward Germany. For the next several months the Allies were compelled to slog their way through forest, cities, and the fortifications of the Westwall.[72]

The tank destroyer battalions never maneuvered as had been envisioned; they were farmed out to the divisions to augment their combat power. The Germans executed only one massed armored attack, which was the very threat that tank destroyers had been created to counter. This massed armored attack occurred during the winter 1944 Ardennes offensive, and the tank destroyer battalions were largely dispersed throughout the 12th Army Group and incapable of executing the coordinated, mass maneuver called for in doctrine.

---

[71] Gabel, "World War II Armor Operations in Europe," in Hofmann and Starry, eds., *From Camp Colt to Desert Storm*, p. 167.

[72] Gabel, "World War II Armor Operations in Europe," in Hofmann and Starry, eds., *From Camp Colt to Desert Storm*, p. 169.

At the lowest tactical levels, the modern definition of maneuver—"systems move to gain positions of advantage against enemy forces"[73]—took on a particular importance because of the lethality and survivability advantage heavy German armor enjoyed over U.S. medium armor. Major General Maurice Rose, commander of the 3d Armored Division, wrote about this problem in a March 1945 letter to General Eisenhower: "It is my personal conviction that the present M4A3 tank is inferior to the German Mark V." He told General Eisenhower that he had seen "projectiles fired by our 75mm and 76mm guns bouncing off the front plate of Mark V tanks at ranges of about 600 yards." U.S. tank crews had to close the range or angle for flank or rear shots, which was not always possible "due to the canalizing of the avenue of approach of both the German and our tank, which did not permit maneuver."[74]

Fire Support. The U.S. Army enjoyed two significant fire support advantages over the German Army during the campaign in Europe: field artillery and air power. During the period between the two world wars, the U.S. Army had developed fire direction procedures that enabled multiple units to mass their fires on targets. Both air and ground forward observers who were linked to firing units through radios or wire communications provided responsive fires. Self-propelled and motorized field artillery provided a much higher degree of mobility than had been the case in previous wars.[75] Additionally, organizational arrangements, most notably the artillery group,

> permitted commanders to move artillery battalions from army to army, corps to corps, or division to division with ease and furnish additional artillery support where it was needed.[76]

---

[73] U.S. Department of the Army, FM 3-0, *Operations*, p. 5-16.

[74] Major General Maurice Rose to General Dwight D. Eisenhower, March 21, 1945, cited in Johnson, *Fast Tanks and Heavy Bombers*, p. 199.

[75] Dastrup, *King of Battle*, p. 226.

[76] Dastrup, *King of Battle*, p. 220.

The U.S. air-ground system was largely developed during combat operations. The U.S. Army Air Forces had played a key but largely independent role in strategic bombing and had made the invasion possible by isolating the Normandy beachhead.[77] Nevertheless, there was no system in place for supporting ground forces with air power when the invasion occurred.

The system of air-ground cooperation rapidly evolved after the invasion, largely through the initiative of Major General Elwood "Pete" Quesada, commander of the IX Tactical Air Command, which was supporting General Bradley's First Army. General Quesada collocated his operations center with that of Bradley's First Army,[78] and his efforts steadily paid dividends. During its pursuit following the Normandy breakout, the 4th Armored Division availed itself of this evolving system:

> Ninth Air Force's XIX Tactical Air Command provided a four-ship armed reconnaissance flight over each column. The fighter-bomber pilots, who were in direct radio communication with the commanders of the 4th Armored's combat commands, warned the ground elements of obstacles and enemy strong points, many of which they were able to neutralize before the armored columns arrived.[79]

Air power also helped make up for the disparity between German and U.S. armored vehicles. General Rose told General Eisenhower in March 1945 that U.S. soldiers compensated for their "inferior equipment by the efficient use of artillery, air support, and maneuver."[80] Sergeant Harold E. Fulton perhaps said it best: "Our best tank weapon, and the boy that has saved us so many times, is the P-47 [fighter

---

[77] Zetterling, *Normandy 1944*, p. 112.

[78] Hughes, *Over Lord*, pp. 156–158.

[79] Gabel, "World War II Armor Operations in Europe," in Hofmann and Starry, eds., *From Camp Colt to Desert Storm*, p. 169.

[80] General Rose to General Eisenhower, cited in Johnson, *Fast Tanks and Heavy Bombers*, p. 199.

airplane]."[81] At the end of the war, the after-action review of the 12th Army Group noted emphatically the power of integrated ground and air power: "The air-armor team is a most powerful combination in the breakthrough and exploitation. . . . The use of this coordinated force, in combat, should be habitual."[82]

Air Defense. Except in isolated instances, "overwhelming Allied air superiority made an integrated air defense system increasingly unimportant during 1944–1945."[83] Consequently, U.S. antiaircraft units were dispersed and frequently used by commanders in an effort "to engage targets on the ground."[84] In the fall of 1944, many antiaircraft units were inactivated and used "to provide much-needed infantry replacements."[85]

Command and Control. Radio communications played an increasingly important role in World War II. Tactical communications, in particular, were a C2 breakthrough in that they "were the basis for controlling fluid, mechanized operations as well as the raw material for tactical SIGINT."[86]

The U.S. Army employed frequency modulation radios for tactical, short-range communications, and used very high frequency

---

[81] Brigadier General I. D. White to General Eisenhower, March 20, 1945, cited in Johnson, *Fast Tanks and Heavy Bombers*, p. 199.

[82] U.S. 12th Army Group, *12th Army Group Report of Operations*, p. 61. See also The General Board, United States Forces, European Theater, *The Tactical Air Force in the European Theater of Operations*, Foreword, p. 1. This study notes the importance of air power to ground operations: "The entrance of ground components was timed on air capabilities. Air success was so important that if delays occurred due to weather, or other causes, ground action on an Army or larger scale was postponed." However, War Department doctrine was, in the view of this report, inadequate in the realm of ground-air doctrine, and procedures were largely ad hoc: "No manual existed except Field Manual 31-35 . . . which had been obsolete for years. The splendid cooperation between the Tactical Air Commands and the Armies was developed during operations."

[83] House, *Combined Arms Warfare*, p. 168. House believes that the most notable exception was the protection of the Remagen bridge over the Rhine River.

[84] House, *Combined Arms Warfare*, p. 168.

[85] House, *Combined Arms Warfare*, p. 168.

[86] House, *Combined Arms Warfare*, p. 157.

(VHF) and ultra high frequency (UHF) radios for longer-range communications. Radio communications allowed senior commanders to directly monitor operations in real time through tactical frequencies or reports from liaison teams. (Without radios, commanders had to wait for reports to make their way through intervening echelons of command.[87]) The more-direct monitoring system was a two-edged sword, however:

> The danger with such a monitoring system, as Gen. Dwight D. Eisenhower acknowledged after the war, was that the senior commander might be tempted to bypass the intermediate headquarters and interfere directly in the battle, using the system for command rather than as a source of timely operational and intelligence information. In the latter role these monitoring services enabled much more effective coordination of the battle, allowing the commander to react through his subordinate commanders to situations as they developed.[88]

During the exploitation that followed Operation Cobra, radio communications were crucial. The actions of Major General John S. Wood, commander of the 4th Armored Division, provide a notable example: "[E]xercising command and control from a liaison aircraft and issuing mission-type orders verbally or over the radio, Wood drove his division more than a thousand road miles in thirty-five days."[89] Furthermore, aircraft communicating via radio to General Wood's commanders provided reconnaissance and air support.[90] Thus, as historian Steven Zaloga notes, radio communications were

> an essential element in modern combined arms warfare. They allowed the division headquarters to coordinate units scattered

---

[87] House, *Combined Arms Warfare*, pp. 157–158.

[88] House, *Combined Arms Warfare*, p. 158

[89] Gabel, "World War II Armor Operations in Europe," in Hofmann and Starry, eds., *From Camp Colt to Desert Storm*, p. 167.

[90] Gabel, "World War II Armor Operations in Europe," in Hofmann and Starry, eds., *From Camp Colt to Desert Storm*, p. 169.

> over dozens of miles, even when all the units were in motion. At a tactical level, they permitted the coordination of different combat arms in real time, increasing the flexibility of units on the battlefield. Radio substantially enhanced the effectiveness of artillery since forward observers could call down fire with considerable precision precisely when it was needed. Artillery fire is far more lethal when used in a directed fashion rather than in preplanned fire strikes, so radio was instrumental in substantially increasing the lethality of artillery. Likewise, innovations in tank-mounted radios in 1944 made it possible for armored columns to coordinate air strikes from roving fighter bombers. Radio had a much greater impact on US armored tactics than in other armies as radios were more widely distributed, and of a significantly better quality.[91]

Significant communications interoperability issues did surface during operations, however. To begin with, "the radios issued to infantry, tank, and fighter aircraft units had different frequency spectra, making communications among the arms impossible."[92] Improvisations in the field—ranging from mounting external telephones on tanks to communicate with infantry and installing VHF aircraft radios on tanks to allow air controllers to call in air strikes—solved some of these issues, and better enabled combined-arms warfare.[93] The important point is that problems were not anticipated before operations; thus, solutions generally came from below, not from the War Department.

Combat Service Support. Armored warfare requires enormous logistical support. This became excruciatingly apparent as U.S. supply lines from the Normandy beachhead were stretched almost to the breaking point during the pursuit after Operation Cobra. As previously noted, logistical overextension was one of the major contributors

---

[91] Steven J. Zaloga, *US Armored Divisions: The European Theater of Operations*, University Park, Ill.: Osprey Publishing, 2004, p. 48.

[92] House, *Combined Arms Warfare*, p. 167.

[93] House, *Combined Arms Warfare*, p. 167.

to the stalling of the pursuit: "The armored division lived on oil and gasoline."[94]

Furthermore, stresses on the support system resulted in rationing of all classes of supply, including artillery ammunition.[95] In consequence, "staffs were taught that existing tables of organization and equipment had to be augmented with extra men and twice the number of supplies."[96]

Characteristics of Transformation Insights. This case offers insights in deployability, versatility, lethality, and survivability.

Deployability. Deployability was a key consideration for the U.S. Army because the theaters of war were distant from the United States. General Marshall discussed this focus on deployability—and its consequences—in his 1945 report to the Secretary of War:

> Another noteworthy example of German superiority was in the heavy tank. From the summer of 1943 to the spring of 1945 the German Tiger and Panther tanks outmatched our Sherman tanks in direct combat. This stemmed largely from different concepts of armored warfare held by us and the Germans, and the radical difference in our approach to the battlefield. Our tanks had to be shipped thousands of miles overseas and landed on hostile shores amphibiously. They had to cross innumerable rivers on temporary bridges, since when we attacked we sought to destroy the permanent bridges behind the enemy lines from the air. Those that we missed were destroyed by the enemy when he retreated. Therefore our tanks could not well be of the heavy type. We designed our armor as a weapon of exploitation. In other words, we desired to use our tanks in long-range thrusts deep into the enemy's rear where they could chew up his supply installations and communications. This required great endurance—low consumption of gasoline and ability to move great distances without breakdown.

---

[94] Jarymowycz, *Tank Tactics*, p. 216.

[95] Dastrup, *King of Battle*, p. 226.

[96] Jarymowycz, *Tank Tactics*, p. 216.

> But while that was the most profitable use of the tank, it became unavoidable in stagnant prepared-line fighting to escape tank-to-tank battles. In this combat, our medium tank was at a disadvantage, when forced into a head-on engagement with the German heavies.[97]

General Marshall's deployability argument was not just a post-war rationalization. However, the comments of Brigadier General J. H. Collier, commander of Combat Command A, 2nd Armored Division, suggest that General Marshall's justification fell flat with soldiers even during the war:

> The fact that our equipment must be shipped over long distances does not, in the opinion of our tankers, justify our inferiority. The M4 has proven inferior to the German Mark VI in Africa before the invasion of Sicily, 10 July 1943.
>
> It is my opinion that press reports by high ranking officers to the effect that we have the best equipment in the world do much to discourage the soldier who is using equipment that he knows to be inferior to the enemy.[98]

Baily offers a balanced assessment of the trade-offs apparent in the fielding of inferior U.S. armored vehicles during World War II, and of the ensuing results:

> The lack of concern about tank development may have been justified. After all, the US Army never suffered a major tactical reverse because of the quality of its tanks or tanks destroyers. The Shermans and M10s usually had numbers, airpower, and superior

---

[97] U.S. War Department, *Biennial Report of the Chief of Staff of the U.S. Army, General George C. Marshall, July 1, 1943, to June 30, 1945, to the Secretary of War*, Washington, D.C.: U.S. News Publishing Corp., 1945.

[98] General White to General Eisenhower, cited in Johnson, *Fast Tanks and Heavy Bombers*, pp. 197–198. See also Green et al., *The Ordnance Department*, p. 278, which notes that "Army Regulations 850-15 . . . prescribed that no tank weigh more than 30 tons or exceed 103 inches in width, though as one [U.S. Army] Ordnance tank specialist observed, Hitler's tanks violated this American rule."

> artillery. . . . We won the war with the M4. But in northwest Europe in 1944 and 1945, particularly in the snow-clad forests of the Ardennes, the American citizens who manned the Army had to pay in blood for the US Army's failure to provide them with better weapons.[99]

Lethality. U.S. tanks and tank destroyers were at a distinct disadvantage in direct fire engagements with most German tanks and tank destroyers. Tables 2.2 and 2.3 show the ranges at which German and U.S. armor could attack each other with penetrating shots. Quite simply, U.S. tanks had not been designed with tank-on-tank combat in mind. Thus, they had to close the range in order to effectively engage German armored vehicles. A preferred U.S. tactic, first learned in North Africa, was to maneuver into a position that allowed shots to be aimed at the less heavily armored sides and rear of German tanks. The Germans adapted to this tactic, however, and sited their tanks where the less vulnerable fronts of their tanks faced the most likely avenue of approach.

The postwar 12th Army Group after-action report made several telling observations about the lethality of future U.S. armor design:

> [First, our] old tendency to build guns with a long tube life and a consequent necessity for keeping the muzzle velocity low should be studied with caution. The life of a tank in combat is short. It is more economical to change gun tubes than it is to replace tanks KO'd in an unequal gun duel. . . .
>
> [Second, careful] consideration should be given to the abolition of the tank destroyer units as such. The tank has proven to be the best tank destroyer. . . . [I]n spite of pre-war doctrine, the "Seek and Destroy" role against enemy armor is best performed by the tank.[100]

---

[99] Baily, *Faint Praise*, p. 146.

[100] U.S. 12th Army Group, *12th Army Group Report of Operations,* p. 60.

A General Board convened at the end of the war assessed U.S. Army performance in the European theater and judged existing tanks inadequate to execute U.S. armored doctrine. Fundamental to the board's analysis of U.S. armored vehicle performance was the premise that the "European campaign demonstrated that tanks fight tanks."[101] This led to the recommendation that the U.S. Army adopt as the

> *minimum* standard for future [tank gun] development . . . [f]or exploitation tanks of an armored division, a "gun capable of penetrating the sides and rear of any enemy armored vehicle and the front of any but the heaviest assault tank," at normal tank fighting ranges.[102]

Even reconnaissance (light) tanks required a gun that was able to penetrate the "sides and rear of any enemy armored vehicle."[103]

Acknowledgement that "tanks fight tanks" also signaled the beginning of the end of tank destroyers in the U.S. Army. Armored division commanders believed that, given "the trend to tanks with high velocity weapons capable of destroying other tanks," there was no need for tank destroyer units in the armored division.[104] Infantry division commanders agreed: "If a tank is given to the Infantry with a proper anti-tank gun, the division commanders favor the replacement of the tank destroyer with a tank."[105] The War Department eventually acted on this assessment:

> On 10 November 1945, the Tank Destroyer Center terminated its few remaining activities and, without fanfare, ceased to exist. Officers commissioned in the tank destroyers found themselves

---

[101] The General Board, *Tank Gunnery*, p. 29.

[102] The General Board, *Tank Gunnery*, p. 29. Emphasis in the original.

[103] The General Board, *Tank Gunnery*, p. 29.

[104] The General Board, *Organization, Equipment, and Tactical Employment of the Armored Division*, p. 22.

[105] The General Board, *Organization, Equipment, and Tactical Employment of Separate Tank Battalions*, Appendix 2.

> transferred to the infantry. . . . The very last tanks destroyer battalion . . . was inactivated . . . on 1 November 1946.[106]

Survivability. As Table 2.2 indicates, U.S. tanks and tank destroyers were highly vulnerable—at considerable ranges—to almost all German armored vehicles. U.S. armored vehicles also proved very vulnerable to German crew-served high-velocity guns (of calibers up to 88 mm) and man-portable antitank weapons. The Germans also fielded a variety of very effective man-portable weapons (including the *racketen-panzerbuchse*, the *panzerschreck*, and the ubiquitous *panzerfaust*), some of which were originally a reverse-engineered improvement on U.S. bazookas captured from the Russians in 1942. All of these weapons could penetrate all U.S. armored vehicles at the ranges at which the tanks were designed to be employed.[107]

From D-Day on June 6, 1944, to the end of the war in Europe on May 12, 1945, the 12th Army Group lost 4,095 tanks. This total included 3,139 M4 Sherman tanks (equipped with 75-mm or 76-mm guns).[108] Recognizing the vulnerability of U.S. armored vehicles in the European theater, the General Board made a very specific recommendation: "Frontal armor and armor over ammunition stowage must be capable of withstanding all foreign tank and anti-tank weapons at normal combat ranges."[109]

Complex Terrain Insights. U.S. armored forces that participated in the Normandy, northern France, Rhineland, Ardennes-Alsace, and central Europe campaigns of World War II fought in compartmentalized terrain (e.g., in *bocage* country), urban areas, and forests. U.S. Army doctrine and materiel were found wanting in each of these envi-

---

[106] Christopher R. Gabel, *Seek, Strike, and Destroy: U.S. Army Tank Destroyer Doctrine in World War II*, Fort Leavenworth, Kan.: Combat Studies Institute, U.S. Army Command and General Staff College, 1985, p. 65.

[107] John Weeks, *Men Against Tanks: A History of Anti-Tank Warfare*, New York: Mason/Charter, 1975, pp. 51–69.

[108] U.S. 12th Army Group, *12th Army Group Report of Operations*, p. 66.

[109] The General Board, *Organization, Equipment, and Tactical Employment of Separate Tank Battalions*, p. 12.

ronments. Historian Christopher Gabel provides the following sober assessment:

> Postwar commentators tended to remember the dashing, blitzkrieg-type operations rather than the slugfests in the Normandy hedgerows or the streets of Aachen. They tended to remember the Sherman tank's weaknesses as an antitank system, while forgetting that 70 percent of the rounds fired by the Sherman were high-explosive, not antitank, ammunition. Day in, day out, armor's chief contributions were in functions that armor doctrine said tanks should avoid: fighting in cities, reducing pillboxes, and generally operating at the pace of the infantry. The blitzkriegs stand out precisely because they were the exception rather than the rule.[110]

U.S. Army Armored Cavalry and Mechanized Infantry in Vietnam (1965–1972)

DOTMLPF Insights. This case offers insights in the areas of doctrine, training, leader development, organization, and materiel.

Doctrine. U.S. Army doctrine for the employment of U.S. armored forces focused on conventional combat in Europe.[111] As previously noted, U.S. armored forces largely abandoned these conventional concepts and adapted to the environment in which they found themselves. The U.S. Army Armor School and the Combat Developments Command Armor Agency at Fort Knox, Kentucky, rejected any doctrinal changes based on U.S. experience in Vietnam, arguing that "new concepts were not applicable to armor combat in other parts of the world."[112] Regarding the use of M113s as fighting vehicles, the CONARC believed that "adopting as doctrine the employment of mounted infantry in a cavalry role was neither feasible nor desirable."[113]

---

[110] Gabel, "World War II Armor Operations in Europe," in Hofmann and Starry, eds., *From Camp Colt to Desert Storm*, p. 179.

[111] Starry, *Mounted Combat in Vietnam*, p. 7.

[112] Starry, *Mounted Combat in Vietnam*, p. 86.

[113] Starry, *Mounted Combat in Vietnam*, p. 86.

Training. The U.S. Army Armor School and armored units in the United States focused on training armored formations for Europe. Therefore, the TTP used in Vietnam were learned and employed largely within the combat zone. Indeed, "[a]rmored units arriving early in the Vietnam War literally had to invent tactics and techniques, and then convince the Army that they worked."[114]

Leader Development. The policies of one-year combat tours and of assigning officers to leadership positions (i.e., platoon leader, company or troop or battery commander, battalion or squadron commander) for only six months negatively affected both leader development and unit cohesion.[115]

Organization. Armored units were frequently employed in a piecemeal fashion to support infantry. This practice resulted in few operational penalties, given the absence of enemy armor, but it did create significant logistical difficulties. Armored units detached from their parent organization frequently suffered maintenance and supply difficulties.[116]

Materiel. Units modified M113s and M551s in the field to increase tank survivability and lethality. M113s were modified into ACAVs that mounted machine guns with armor shields. Follow-on modifications

> provided thicker belly armor to protect crews from mine explosions, the relocation and strengthening of the fuel line to lessen the danger of fire, and stand-off shielding designed to cause the premature detonation of the enemy's lethal rocket propelled grenades. . . .[117]

To protect the vehicle commander, M551s received a shield around the turret-mounted .50-caliber machine gun. Nevertheless, some of the

---

[114] Hofmann and Starry, eds., *From Camp Colt to Desert Storm*, p. 65.

[115] Russell W. Glenn, *Reading Athena's Dance Card: Men Against Fire in Vietnam*, Annapolis: Naval Institute Press, 2000, pp. 118–121.

[116] Glenn, *Reading Athena's Dance Card*, p. 221; Hofmann and Starry, eds., *From Camp Colt to Desert Storm*, pp. 336–337.

[117] Hofmann and Starry, eds., *From Camp Colt to Desert Storm*, p. 331.

Sheridan's significant survivability and reliability problems were never completely resolved. Its combustible case ammunition "could be detonated by a mine blast or a hit by a rocket propelled grenade."[118] The Sheridan also suffered

> persistent problems with incomplete combustion of the main gun shell casings and with malfunctions of the electrical firing system, especially in wet weather. A number of common difficulties with the system's durability, such as overheated engines, turret electrical power failures, and failure of the gun's recoil system were also encountered.[119]

Finally, the Sheridan's "hot, cramped crew compartment caused significant fatigue."[120]

BOS Insights. This case yields BOS insights in the areas of maneuver, fire support, intelligence, mobility and survivability, and CSS.

Maneuver. The chief impediments to U.S. mounted maneuver were random-mining techniques used by enemy forces and ambushes.

Fire Support. In Vietnam, the U.S. Army employed the massive firepower resources at its disposal. General Harry W. O. Kinnard's perspective on the use of firepower was probably representative of the views of many U.S. officers:

> Kinnard emphasized that he never failed to expend all of his available firepower to support troops in combat. "When you have it, you use it," he replied. "To do otherwise only risks the success of the operation and needlessly gets soldiers killed".[121]

Some officers, however, had a different attitude:

---

[118] Starry, *Mounted Combat in Vietnam*, pp. 144–145.

[119] Hofmann and Starry, eds., *From Camp Colt to Desert Storm*, p. 334.

[120] Hofmann and Starry, eds., *From Camp Colt to Desert Storm*, p. 334.

[121] Robert H. Scales, Jr., *Firepower in Limited War*, Washington, D.C.: National Defense University Press, 1990, p. 247.

> Firepower too easily becomes an acceptable and quick solution for commanders who have neither the experience nor the time to come to grips with the militarily elusive and politically sophisticated challenges of counterinsurgency operations. It is through overemphasis and over-reliance on artillery and aerial bombardment that commanders change effective military tactics into counterproductive operations.[122]

Intelligence. Accurate intelligence in the mixed counterinsurgency and conventional environments that existed simultaneously at different points during the war in Vietnam, while critical to successful operations, was very difficult to attain.

Mobility/Countermobility/Survivability. The principal threat to armored vehicle mobility in Vietnam came from landmines and IEDs. Despite the use of minerollers and landmine detectors, these weapons took a heavy toll. A good example is the experience of the 11th Armored Cavalry Regiment from June 1969 to June 1970. The unit "encountered over 1,100 mines in the northern III Corps Tactical Zone. Only 60 percent were detected; the other 40 percent accounted for the loss of 352 combat vehicles."[123]

Combat Service Support. Two issues conspired to complicate CSS of U.S. armored forces in Vietnam. First, armored units, as previously noted, were often split up and farmed out to various maneuver units. This made their maintenance and supply difficult. Second, the M48A3 tank was an old, second-tier U.S. Army system. One squadron executive officer was frustrated by the need to "piece together our fifteen-year-old tanks as best we could" and "wondered at a policy that sent new tanks to Europe and old tanks with only inadequate repair parts available to the combat zone."[124]

Characteristics of Transformation Insights. This case yields transformation insights in the areas of agility, versatility, lethality, surviv-

---

[122]Charles K. Nulsen, "Advising as a Prelude to Command," essay, Carlisle, Pa.: U.S. Army War College, December 2, 1969, p. 3.

[123]Starry, *Mounted Combat in Vietnam*, p. 79.

[124]Michael D. Mahler, *Tinged in Steel: Armored Cavalry in Vietnam, 1967–69,* Novato, Calif.: Presidio Press, 1986, p. 103.

ability, and sustainability. Note that sustainability insights are also discussed above in the BOS section on CSS.

Agility. Armored forces adapted themselves to a wide variety of roles throughout the Vietnam War. Many of these adaptations, such as using ACAVs as fighting vehicles, developed in response to "local conditions" and were not adopted throughout the U.S. Army. Armored forces executed a wide variety of missions, including offense, defense, convoy escort, search and destroy, and fixing.

Versatility. Armored units were task organized (sometimes in a piecemeal fashion that affected their sustainability) to add combat power to infantry forces. Nevertheless, these units were able to transition effectively between operations and were proficient at using joint enablers (e.g., artillery and close air support).

Lethality. U.S. armored forces had a clear lethality advantage over VC and NVA forces, particularly when armor was coupled with massive firepower. When they were able to find, fix, and engage the elusive enemy—a task that proved difficult—U.S. forces enjoyed a high exchange ratio. The Vietnam War raised legitimate issues about the utility of using massive, and often indiscriminate, firepower in a counterinsurgency.

Survivability. Armored vehicles afforded U.S. forces protected mobility and firepower. The enemy, however, discovered and exploited the vulnerabilities of U.S. armored vehicles, particularly M113s and M551s. These vehicles were very susceptible to mines, IEDs, RPGs, and recoilless-rifle fire, which the communist forces used to great effect. The heavier M48A3 tanks were less vulnerable.[125] Nevertheless, M113s gave mechanized infantry and cavalry units protection from booby traps and antipersonnel mines, thus reducing casualties.[126]

Complex Terrain Insights. U.S. armored forces operated in jungle, semimountainous, and urban terrain.

Jungle and Semi-Mountainous Terrain. The 1967 MACOV study showed that armored forces could in fact operate in the jungle and semimountainous terrain of South Vietnam. M113s, in particular,

---

[125] Hofmann and Starry, eds., *From Camp Colt to Desert Storm*, p. 338.

[126] Starry, *Mounted Combat in Vietnam*, p. 64.

could be used in the majority of the theater of operations with great effect. The MACOV report also noted that "armored cavalry was probably the most cost-effective force on the Vietnam battlefield."[127]

Urban Terrain. Armored forces were used very effectively to respond to communist offensives in 1968. Armored mobility enabled U.S. cavalry, tank, and mechanized infantry forces to respond rapidly to enemy actions in several South Vietnamese cities during these offensives. Additionally, armored vehicles provided protected firepower for U.S. forces engaged in the difficult task of clearing enemy forces from cities they had taken during the offensives, particularly during the intense fighting in Hue.

Task Force Shepherd, 1st Marine Division, in Operation Desert Shield and Operation Desert Storm (Southwest Asia, 1990–1991)

DOTMLPF Insights. This case offers insights in the areas of training and materiel.

Training. The U.S. Marines of Task Force Shepherd were highly trained and competent. Their performance at OP 4, where outnumbered light-armored vehicle companies orchestrated combined-arms attacks to defeat Iraqi armored formations, is the clearest example of their high level of training.

Materiel. General Hopkins believes that Marine LAVs made a significant contribution to U.S. Marine operations in Desert Shield and Desert Storm:

> The 25-mm chain gun was deadly. The LAV held up. It could go 30 to 40 miles per hour across the desert floor. We used it when we were determining where we were going to breach and before G-Day, we used the LAV to run up and down the border of Kuwait to confuse the Iraqis on where our penetration was going.[128]

---

[127]Hofmann and Starry, eds., *From Camp Colt to Desert Storm*, p. 343.

[128]Melon et al., *Anthology and Annotated Bibliography*, p. 32. After the 1st Marine Division was in place, General Hopkins left command of the 7th MEB to assume duties as the deputy commander of I Marine Expeditionary Force.

Additionally, the LAV-AT with its TOW ATGMs and thermal sights proved particularly effective against Iraqi armor. The other LAV variants did not have this capability, however, and

> [p]assive night sights were inadequate because they required more ambient illumination than was always available and because they provided no day or night capability on an obscured battlefield. The LAV-AT, however, was equipped with AN/TAS-4 thermal sights for its TOW missiles, which the Marine Corps called, "the single most significant system enhancement of Operation Desert Shield/Desert Storm." . . . Without thermal imaging, the LAV battalion experienced severe operational restrictions in low visibility conditions. As a result, it was recommended that the LAV be equipped with thermal sights and vision devices.[129]

BOS Insights. This case yields BOS insights in the areas of maneuver and fire support.

Maneuver. The LAV provided Task Force Shepherd with protected mobility compared to U.S. Marine light infantry, who were transported in AAV-7s or cargo trucks. This medium-armored force provided the 1st Marine Division with a highly mobile task force that could screen the division and rapidly reposition itself (compared to U.S. Marine tank units) in response to changing orders.

Fire Support. The engagement at OP 4 demonstrated the devastating effects of combined-arms action by skilled professionals. Artillery and air support enabled the outnumbered forces of Task Force Shepherd to defeat the Iraqi forces that attacked OP 4.

Characteristics of Transformation Insights. This case yields transformation insights in the areas of agility, versatility, lethality, and sustainability.

Agility. Throughout the operation Task Force Shepherd adapted to changes in mission, including screening, reconnaissance, defense, and assaulting objectives.

---

[129]Cordesman and Wagner, *The Gulf War*, p. 705.

Versatility. Task Force Shepherd reorganized on several occasions during the Gulf War. It was able to attach and detach units in accordance with division orders without any apparent difficulty.

Lethality. As previously noted, the LAV-AT (with its TOW ATGM and thermal sights) proved extremely effective as a long-range system against Iraqi armor, particularly during periods of limited visibility. Because Iraqi armor could not engage the LAV-ATs before being destroyed, the LAV-ATs' ability to deliver long-range, stand-off fires afforded them protection. The LAV-25 chain gun system proved highly lethal against light and medium armor.

Sustainability. LAVs proved highly reliable and sustainable.

Complex Terrain Insights. Task Force Shepherd encountered complex terrain in the form of obstacle belts on the Kuwait Border and reduced visibility caused by oil field fires. Because it had no mine-clearing capability of its own, Task Force Shepherd followed the other elements of the 1st Marine Division through the obstacle belt. On one occasion, Lieutenant Colonel Myers followed a set of fresh tire tracks through a minefield.[130] Because the LAV-AT's thermal sights could see through obscurants, the unit was able to operate in the limited-visibility conditions caused by the burning oil fields with less difficulty than other U.S. Marine forces.

Medium-Armored Forces in the Center of the Range of Military Operations

Soviet Airborne Operations in Czechoslovakia (1968)

DOTMLPF Insights. This case offers insights in the areas of doctrine, training, and materiel.

Doctrine. Operation Danube demonstrated the soundness of the Soviet rapid airborne insertion doctrine. These rapidly arriving troops were able to quickly execute a coup de main of the Czechoslovakian government.

[130]Cureton, *United States Marine Corps in the Persian Gulf*, p. 83.

Training. The Soviet airborne units that executed Operation Danube were well trained. Zaloga notes that the VDV's performance made it "the pride of the Soviet Army. . . . In contrast, after action reports castigated the tank and motor rifle forces for their poor levels of preparedness."[131]

Materiel. The medium-armored vehicles used by the VDV were air-landable and gave the paratroopers protected mobility and sufficient firepower to accomplish their missions.

BOS Insights. This case yields BOS insights in the areas of maneuver and intelligence.

Maneuver. VDV forces moved quickly to their objectives in their organic and commandeered vehicles, capitalizing on the element of surprise and a largely passive Czech population to rapidly establish their presence in Prague.

Intelligence. The Soviets had done an excellent job of preparing for Operation Danube. They knew the locations of critical installations and leaders before the operation began. The KGB aided the VDV forces in the execution of their missions.

Characteristics of Transformation Insights. This case yields transformation insights in the areas of responsiveness, deployability, lethality, and survivability.

Responsiveness. VDV forces provided the Soviet Union with a highly responsive force, albeit of limited size, that could rapidly seize key nodes within Prague and execute a take-down of the government.

Deployability. VDV forces were very deployable because of the combination of Antonov An-12 military transports and the VDV division's organic air deployable medium-armored vehicles. These capabilities enabled the Soviets to rapidly build up combat power, maneuver to key installations, and execute a coup de main.

Lethality. The medium-armored vehicles used by the 7th Guard Airborne Division in Prague would have met organized Czech resistance—not to mention Czech MBTs—with difficulty. Their signal achievement was in preempting such resistance, not defeating it.[132]

---

[131] Zaloga, *Inside the Blue Berets*, p. 163.

[132] Zaloga, *Inside the Blue Berets*, p. 150.

Survivability. The medium-armored vehicles used in Prague provided crews with mobility and protection from civilians in the streets. The principal threat to the vehicles, in the absence of the Czechoslovakian Army, came from "adventurous Czech teenagers . . . attempting to set them on fire with oily rags." Soviet crews fired machine guns into the crowds to warn them off.[133]

Complex Terrain Insights. The paratroopers of the 7th Guards Airborne Division operated in a MOUT environment that was highly permissive. Because of the rapidity of the Soviet assault and the quick neutralization of the Czechoslovakian Army, the paratroopers encountered little resistance. The medium-armored vehicles used by the airborne units gave them protected mobility and intimidated the inhabitants of Prague, who did not appear terribly willing to resist anyway.

South Africa in Angola (1975–1988)

DOTMLPF Insights. This case offers insights in the areas of doctrine, training, organization, and materiel.

Doctrine. MPLA and SAA armor doctrine were a study in contrasts. MPLA operations featured massive, multibrigade sweeps by mechanized infantry and armor formations. MPLA brigades tended to move as complete entities, and often bunched together on the available LOCs. This was due to a mix of logistics constraints and a doctrine that kept MPLA ground forces within the surface-to-air missile umbrella (even after the SAAF largely ceded air superiority in the late 1980s to Cuban- and MPLA-piloted MiG-21s, MiG-23s, and SU-22s).[134]

SAA medium-armor units were employed in much smaller, more dispersed packages.[135] In contrast to MPLA's broad offensives, South African offensive operations took the form of deep-penetration, blitzkrieg-type operations.[136] For example, South Africa opened its conven-

---

[133] Zaloga, *Inside the Blue Berets*, p. 162.

[134] Heitman, "Operations Moduler and Hooper."

[135] Bridgland, *The War for Africa*, pp. 55–70.

[136] Bridgland, *The War for Africa*, pp. 17, 64. Indeed, Soviet advisors planned many MPLA operations. General Konstantin Shaganovitch assumed command of the combined MPLA, Cuban, Soviet, and East German forces in Angola in 1985.

tional involvement in Angola with Operation Savannah, which featured offensive operations conducted by three dispersed mechanized battalion task forces. The task forces penetrated deeply into Angola to induce operational paralysis in MPLA forces and threaten the MPLA government with a coup de main. In a matter of weeks, South African pincers almost reached to the Angolan capital of Luanda—some 3,000 km from the border.[137] There is little doubt that the South Africans would have seized Luanda had the United States not intervened to prevent a major superpower crisis.

SAA operations also relied heavily on the thorough integration of special forces operators, called "recces." The SAA has a world-renowned special operations capability resident in its 4th Recce Regiment. During the war in Angola, the regiment conducted independent operations in support of conventional South African forces.[138] One famous example is the recce's shadowing of the key MPLA airbase in southern Angola. After the MPLA fielded MiG-23s to the region, the SAAF lost the ability to contest air superiority. South Africa's response was to infiltrate recces to locations very near the airbase; from these locations, the recces reported departures and arrivals. When MiG-23s were launched, South African air operations were shaped to avoid interception. When ground attack aircraft were launched, South African ground operations—such as fire support and counterbattery operations by G-5 batteries—were shifted to avoid attack until the threat had receded. This unique recce mission, which was sustained for some time, allowed SAAF and SAA operations to proceed without effective intervention from MPLA aircraft.[139]

In addition to independent missions, recce teams were integrated into South African battalion task forces.[140] The recces were absolutely vital to South Africa's success on the ground. They were perhaps the most valuable source of intelligence for South African commanders, as they

---

[137] Steenkamp, *Borderstrike!*, p. 3.

[138] Bridgland, *The War for Africa*, p. 40.

[139] Bridgland, *The War for Africa*, p. 67.

[140] Lord, "Operation Askari."

operated deeply and covertly behind MPLA lines. Often establishing their observation posts only meters from MPLA positions, recce teams shadowed MPLA units and continually reported on MPLA strength and location. Additionally, the recces frequently called in bombardments from South African artillery batteries up to 40 km away, causing casualties among MPLA units and eroding their morale.[141]

Another key aspect of SAA operations was the very aggressive use of artillery. The South Africans were fortunate to have, in the G-5, perhaps the finest tube artillery piece in the world.[142] With a range of more than 40 km and extraordinarily lethal airburst 155-mm ammunition, the G-5 may have caused more MPLA casualties than any other South African weapon.[143] The SAA distributed G-5 batteries to battalion task forces operating in Angola, and supported the G-5 batteries with additional batteries of 81-mm mortars and 127-mm free-flight rockets.[144] Utilizing recce spotters, SAAF liaison teams accompanying friendly ground forces, spotter aircraft, and RPVs, used forward-deployed G-5 batteries to pound MPLA positions constantly and to great effect.[145] This bombardment was particularly effective due to the extraordinary accuracy of the G-5, which allowed spotters to target and destroy individual vehicles and fighting positions with very few "walk-in" rounds.[146] Spotter adjustments of a few meters were not uncommon.

Training. In this period, the SAA provided world-class training to all ranks, from enlisted to senior officers. The uncompromising training standards set for all tasks—from tactical drills to campaign planning—provided a (perhaps *the*) decisive advantage over Angolan and Cuban forces. One historian observed the following:

---

[141] Heitman, *War in Angola*, p. 343.

[142] Bridgland, *The War for Africa*, p. 87.

[143] Heitman, *War in Angola*, p. 344.

[144] Heitman, *War in Angola*, p. 343.

[145] Bridgland, *The War for Africa*, p. 115.

[146] Bridgland, *The War for Africa*, p. 87.

> Perhaps the most important conclusion to be drawn from the operations of this campaign is that the South African Army has a training programme that works and that produces some excellent soldiers and junior leaders. The proof is to be found in the very low casualties suffered in the campaign. The heaviest fighting involved mechanized clashes at short ranges in very close terrain. Any weaknesses in either the training or the placing of soldiers and junior leaders would have resulted in heavy casualties.[147]

SAA training was not, however, without its weak points. Much of the fighting in Angola was accomplished by elements of the Southwest Africa Task Force or indigenous forces that were not as well trained as conventional SAA units. Furthermore, the SAA began to rely on units staffed with national service conscripts as the war widened. While these national servicemen generally performed quite well, their annual rotation frustrated commanders and forced operations to be conducted around the arrival of new levies.[148] Additionally, South African armor crews were normally trained on 1,000-meter ranges, a stark contrast to the environment in Angola that routinely produced engagement ranges under 20 meters.[149]

Despite these weaknesses, South Africa fielded by far the best soldiers of any of the belligerents. Virtually every history of the war refers to the SAA's superiority in field craft, discipline, and planning. In many cases, superior training allowed SAA units to meet and defeat MPLA units that were equipped with more-lethal and more-survivable vehicles.

Organization. During the Border War, both MPLA and SAA units fought as combined-arms units. As previously noted, the MPLA tended to fight as brigades while the SAA fielded large, task-organized "combat groups" commanded by an infantry battalion headquarters. South African task forces were provided with a remarkable array of capabilities, including infantry, armor, mortars, tube artillery, rocket

---

[147] Heitman, *War in Angola*, p. 342.

[148] Bridgland, *The War for Africa*, p. 225.

[149] Bridgland, *The War for Africa*, p. 227.

artillery, special forces, electronic warfare (EW), and logistics elements. These capabilities allowed the South Africans to achieve the survivability and lethality levels offered by combined-arms formations while remaining sustainable over extremely difficult LOCs.

Materiel. SAA operations in Angola yield a number of materiel lessons. SAA mechanized vehicles favored mobility over armor, subscribing to the theory that more-mobile forces could decline engagements with heavier and more-lethal adversaries, choosing instead to maneuver into advantageous engagements. However, the close terrain of southern Angola obliged SAA mechanized units to accept very short-range engagements, typically a distance of between 20 and 200 meters.[150] This remained true despite clear South African superiority in what today would be called C4ISR, including the advantages they accrued from SAAF 10 Squadron Seeker RPVs and outstanding support from the recce teams who operated independently and in direct support of the task forces.[151] Survivability of South African mechanized vehicles remained a concern throughout the Border War.

The South African experience in Angola also suggests that survivability involves more than simply neutralizing the effects of enemy action. The SAA found that the survivability problems presented by terrain and the environment present were just as important as the effects of enemy action. One senior SAA officer wrote that

> [t]he bush war taught us that AFVs need to be extensively armoured, not only against enemy projectiles but also against the damage caused by vegetation. Vehicles must have smooth and hardened exteriors, with no protruding objects and no exposed components (especially under the chassis). Essential protrusions such as antennae, lights, etc. must be specially protected. Adequate climate control (ventilation/air conditioning) and availability of water for closed-down tank crews is essential to prevent dehydration and heat fatigue.[152]

---

[150] Bridgland, *The War for Africa*, p. 80.

[151] Heitman, *War in Angola*, p. 344.

[152] J. M. Dippenaar, "Armour in the African Environment," in Jakkie Cilliers and Bill Sass, eds., *Mailed Fist: Developments in Modern Armour*, Institute for Security Studies, Monograph

As a result of this lesson, modern South African mechanized vehicles are very large and heavy in spite of their wheels. Their high profiles provide sufficient ground clearance and mine survivability.

The MPLA's lackluster performance is also instructive. Poor performance has been blamed on the low serviceability rates of the electronic components on late-model Soviet vehicles that were subject to brutal "bundu-bashing" as forces advanced and retreated across the rough African terrain.[153]

Lethality was also a major materiel issue. South African mechanized vehicles lacked lethality against MPLA vehicles, particularly late-model Soviet MBTs. Much like U.S. forces in 1944–1945 on the Western Front, the South Africans found that their Ratels required four or five rounds from their 90-mm main armament to disable MPLA MBTs. This forced the Ratels to operate as platoons whenever MBTs were present.[154] As a result, the South Africans expended great energy in fielding Olifant MBTs, which were armed with the 105-mm L7 main gun. The Olifants, however, arrived too late and in too limited numbers to make a major mark on the campaign.

The South African G-5 combined an excellent gun with highly lethal airburst ammunition. The G-5 probably inflicted more damage on MPLA forces, including MBTs and other AFVs, than any other system.[155]

BOS Insights. This case yields BOS insights in the areas of maneuver, fire support, air defense, C2, mobility and survivability, and CSS.

Maneuver. The SAA's experience in Angola suggests that highly dispersed mechanized forces can operate against linearly-employed heavy adversaries. Flexibility and deep maneuver can compensate for disadvantages in raw lethality and survivability, provided that a favorable and significant imbalance in situational awareness exists. Dispersed

---

No. 2, March 1996.

[153] Heitman, *War in Angola*, p. 327.

[154] Bridgland, *The War for Africa*, p. 139.

[155] Heitman, *War in Angola*, p. 344.

operations can deliver decisive results at the tactical, operational, and strategic levels.

South African experience also suggests that successful dispersed maneuver requires the integration of combined-arms elements at very low levels. This provides the suite of capabilities required by tactical commanders while keeping dispersed forces small enough to be supported by extended LOCs.

Fire Support. Effective fire support was critical to South African success in Angola. The outstanding performance of the South African artillery was made possible by the materiel superiority of its systems, major investments in targeting and acquisition capabilities (e.g., recce teams, RPVs, spotter aircraft, and SAAF liaison teams), and the ability to task organize artillery systems (tube, rocket, and mortar) at very low levels. The prominent role played by South African artillery had the concomitant effect of requiring massive logistical efforts to keep forces supplied in a dispersed theater.

Air Defense. The introduction of advanced Soviet aircraft into the theater in the late 1980s threatened to nullify many SAA advantages, including the important ability of G-5 batteries to operate without impediment. The SAA and SAAF lacked adequate air superiority and air-defense assets to counter advanced Soviet aircraft. In response, the South Africans instituted a number of innovative solutions, including recce "shadowing" of MPLA airbases, long-range bombardment of flight lines and hangars, and the introduction of the U.S. Stinger man-portable air defense system (MANPADS) that the United States provided to UNITA. Nevertheless, this case suggests that a medium-weight force (or any other ground force) will be hard-pressed to maintain its operational effectiveness without air superiority, even in complex terrain.

Command and Control. The SAA's dispersed operations in Angola relied heavily on effective command, control, and communications (C3). While the physical environment presented a challenge to effective C3, South African forces consistently found solutions. The SAA also enjoyed EW superiority over the MPLA and its supporters. At the strategic level, EW and SIGINT were critical to South Africa's overall understanding of its adversary's intent.[156] At the operational level, EW often provided key advantages. In September 1987, for example, South African EW teams detected planning for a major MPLA air and artillery bombardment on South African positions. Taking a page from German practice on the Eastern Front in 1944–1945, Colonel Deion Ferreira withdrew his forces from the impact zone before the beginning of the bombardment and rushed them back into their fighting positions at its conclusion. The South African forces subsequently trounced the attack by the MPLA's 21 Brigade.[157] At the tactical level, South African EW capabilities may have been used to jam MPLA intervehicle communications at the moment of attack, and to adjust artillery fire.[158]

Mobility/Countermobility/Survivability. Mobility played a major role in the Angolan Border War. SAA forces were designed to maximize mobility. Their vehicles prized mobility above other design factors, and their operational scheme of maneuver emphasized mobile task forces operating from dispersed positions across the theater. As described below, this placed heavy logistical burdens on the SAA support structure. It also required South African vehicles to be hardened against environmental hazards. The largely unheralded efforts of SAA engineers were vital to maintaining force mobility. Their extraordinary demining and bridging operations are only the most prominent examples.

Combat Service Support. The Border War presented severe CSS challenges to South African forces. The war was fought by dispersed task forces of SAA units supporting UNITA elements. The task forces were operating over extremely difficult terrain characterized by heavy vegetation, sandy and muddy soil, and rough topography. Much of this terrain was either unmapped or unreliably mapped. The few roads that existed were merely dilapidated remnants of Portuguese colonial macadam. The war was therefore fought around airstrips.[159] The key

[156] Heitman, *War in Angola*, p. 346.

[157] Bridgland, *The War for Africa*, pp. 69–70.

[158] Bridgland, *The War for Africa*, p. 41.

[159] Bridgland, *The War for Africa*, p. 27.

LOCs led from South African territory to airstrips in Angola via SAAF C-130 and C-160 transports, then by helicopter and ground vehicle "forward" to the combat elements.[160]

The dispersed nature of the war and the tenuous LOCs posed immense problems for South African logistics support (particularly consumables) and repair.[161] The rough terrain threw off planning factors for fuel consumption, repair rates, and ammunition resupply.[162] Additionally, South African planners found that ground LOCs were very hard on logistics vehicles.[163] Both South African forces and their adversaries experienced significant CSS problems, which often caused offensives to halt at the end of either side's logistical tether.[164]

Characteristics of Transformation Insights. This case yields transformation insights in the areas of responsiveness, deployability, lethality, survivability, and sustainability.

Responsiveness. The SAA's combination of doctrine, training, organization, and materiel enabled its mechanized forces to be extremely responsive to the needs of South African national leadership. These mechanized forces deployed hundreds or thousands of kilometers to the theater of operations, at times by C-130, and operated with decisive effect against adversaries armed with more-survivable and more-lethal vehicles. They cooperated effectively with UNITA, a nongovernmental coalition ally, and successfully melded ground, air, and special forces into an effective and agile tool.[165]

Deployability. The SAA's medium-weight armor units proved highly deployable. The wheeled Elands, Ratels, Casspirs, and Buffels were often deployed to the theater, and redeployed within it, by C-130 Hercules and C-160 Transall airlifters. The dependability of the wheeled vehicles also allowed SAA units to deploy considerable distances by

---

[160] Stiff, *The Silent* War, pp. 182–234.

[161] Bridgland, *The War for Africa*, p. 73.

[162] Heitman, *War in Angola*, p. 346.

[163] Bridgland, *The War for Africa*, p. 175.

[164] Bridgland, *The War for Africa*, p. 64.

[165] Heitman, "Operations Moduler and Hooper."

surface movement. These attributes created useful options for South African commanders and political authorities in Pretoria.

Operation Savannah provides an excellent example. South Africa entered the war in Angola by executing a vertical envelopment on MPLA forces attacking UNITA positions in southeast Angola. Operation Savannah required the deployment of squadrons of Elands to the remote South African airbase at Rundu, then further airlifting them more than 700 km over the border to the UNITA base in Nova Lisboa, Angola.[166] South African combat groups subsequently penetrated deep into MPLA territory, nearly reaching the capital of Luanda, and unhinged the MPLA's operations in the southeast.

Lethality. South African mechanized forces lacked intrinsic lethality. They compensated for this weakness by effectively integrating indirect-fire lethality into their operations. When direct clashes with MPLA heavy armor were unavoidable, South African mechanized forces performed effectively due to the superiority of their training. One senior South African officer noted the following:

> All these limiting factors on the main armament require armour crews to ensure first-shot hits. This means good reaction times and slick fire orders and drills, which are only possible with superior training[,] the latter being the reason why our armoured crews survived the armoured battles and gained the edge over their Angolan and Cuban counterparts.[167]

Eventually, however, superior training proved insufficient to compensate for the survivability and lethality deficiencies inherent in South African medium-armored vehicles.

Survivability. The South African experience in Angola indicates that situational awareness is absolutely vital to the survivability of a medium-weight force against heavier forces.[168] South African forces

---

[166] Bridgland, *The War for Africa*, p. 8.

[167] Dippenaar, "Armour in the African Environment," in Cilliers and Sass, eds., *Mailed Fist*.

[168] Dippenaar, "Armour in the African Environment," in Cilliers and Sass, eds., *Mailed Fist*.

melded reconnaissance elements and technical systems (e.g., RPVs, EW, and spotter aircraft) to achieve dominant battlespace awareness. As a result, South African forces were frequently able to inflict damage, often via indirect fire, while avoiding direct combat. Nevertheless, even the superiority of South African situational awareness in Angola did not allow South African forces to avoid meeting engagements altogether, especially in the close terrain characteristic of much of the operational area. Eventually, South Africa fielded its own MBT.

Sustainability. Dispersed operations imposed major logistical burdens on the South Africans. While South African mechanized units were able to operate successfully, some officers in the SAA left the campaign believing that spartan logistical arrangements are inappropriate for operations in difficult environments:

> Moving logistical and combat support vehicles over difficult terrain is more difficult than moving AFVs. Logistical vehicles struggle to keep up with the fighting force, and furthermore the vehicles and their cargoes are severely damaged by the vegetation. In the African environment, operations should never be conducted with sparingly deployed logistics, far from depots or third-line support."[169]

Complex Terrain Insights. The operational area in southern Angola was close, and characterized by dense, brushy vegetation, tall grasses, and arid conditions. Furthermore, the majority of Angola's infrastructure was poorly developed. South African operations coped with the close terrain by task organizing down to the tactical level so that the required capabilities would always be close at hand. They also invested heavily in situational awareness, which afforded them some measure of lethality and survivability. Although the close terrain, coupled with the dispersion of South African forces, severely strained South African sustainment capabilities and often limited operations, it probably represented a net advantage for the South African medium-weight force. The terrain limited the mobility of the MPLA forces, limited the long-

---

[169] Dippenaar, "Armour in the African Environment," in Cilliers and Sass, eds., *Mailed Fist.*

range lethality of MPLA MBTs, and created significant sustainment problems for the MPLA.

Soviet Operations in Afghanistan (1979–1989)

DOTMLPF Insights. This case offers insights in the areas of doctrine, training, materiel, and personnel.

Doctrine. The Soviet doctrine for a coup de main against the Afghan government was, initially, highly successful. The Afghan government was rapidly decapitated and replaced with a Soviet puppet regime. In the aftermath of the takeover, however, Soviet doctrine proved highly ineffective at dealing with the Mujahideen insurgency. Although the Soviets adapted their doctrine, the forces available to conduct the counterinsurgency were never sufficient to attain success.

Training. Initially, Soviet soldiers were not adequately trained for service in Afghanistan. Although training improved to some extent as the war went on, the performance of conventional units was often hampered by the quality of their soldiers, who received "relatively little training beyond boot camp."[170] Soviet elite forces (*Spetsnaz*, reconnaissance troops, airborne troops, and helicopter-borne assault troops) were "subjected to rigorous specialized training" and were far more capable than conventional units.[171] Consequently, they were used more often in combat operations, while conventional infantry forces were consigned to garrison, support, and convoy duties.[172]

Materiel. The Soviets learned that lighter armored vehicles were more useful in counterinsurgency, particularly in mountainous terrain. They began to rely more heavily on BMDs for service in the mountains because tanks and BMPs "were too large for mountain trails, subject to frequent breakdowns, and difficult to service in the field."[173] Larger

---

[170] Alexiev, *Inside the Soviet Army in Afghanistan*, p. vi.

[171] Alexiev, *Inside the Soviet Army in Afghanistan*, p. vi.

[172] Cordesman and Wagner, *The Afghan and Falklands Conflicts*, pp. 135–137.

[173] Cordesman and Wagner, *The Afghan and Falklands Conflicts*, p. 150. See also Zaloga, *Inside the Blue Berets*, p. 241. Zaloga notes the following:

> The BMD-1 airborne assault vehicle proved a disappointment in combat. . . . Its suspension had been designed for light weight and as a result was very fragile. It soon became

vehicles, including tanks, BMPs, BTR-60s, BTR-70s, and BTR-80s, were used "in the cities, for LOC protection, as armored ambulances, and in defending strategic crossroads."[174] The Soviets also modified BMDs, BMPs, and other vehicles to make them more effective against infantry targets than against the armored vehicles of NATO that they had been designed to fight. Rapid-fire, hyperelevating 30-mm cannons replaced the 76-mm guns on BMDs and BMPs. Other vehicles mounted hyperelevating machine guns or AGS-17 rapid-fire grenade launchers. These new weapons provided

> machine guns and cannon which could "hose" targets with high rates of fire, easily track targets (unlike slow-moving heavy guns), and be easily hyper-elevated . . . to provide rapid surge fire and direct fire support.[175]

Finally, vehicles often deployed smoke canisters to screen themselves when ambushed.[176]

Personnel. The issue of poor soldier quality in conventional units plagued the Soviets throughout the war in Afghanistan. The Soviet conscription system was corrupt and "conscripts were often sent to Afghanistan because they failed to pay bribes."[177] Furthermore, the regular and support forces included "conscripts with criminal backgrounds and violators of military regulations" who were "sent to Afghanistan in lieu of court-martial."[178] Morale was often low, and there were "serious disciplinary problems, hostility between new and older troops, theft,

---

> chewed up in Afghanistan's rocky terrain. It was too cramped for sustained operations, and its 73mm Grom low-pressure gun could not elevate enough to reach the [M]ujahideen high in the mountains. Its small size made it very vulnerable to mine damage, and the paratroopers soon learned the Vietnam lesson that it was safer to ride outside of a vehicle than inside when mines were present.

[174] Cordesman and Wagner, *The Afghan and Falklands Conflicts*, p. 150.

[175] Cordesman and Wagner, *The Afghan and Falklands Conflicts*, p. 151.

[176] Cordesman and Wagner, *The Afghan and Falklands Conflicts*, p. 153.

[177] Cordesman and Wagner, *The Afghan and Falklands Conflicts*, p. 135.

[178] Alexiev, *Inside the Soviet Army in Afghanistan*, p. iv.

and problems with looting and atrocities." Officers and noncommissioned officers caused problems as well:

> Some officers were, however, sent to Afghanistan as punishment, and many Soviet units seem to have experienced problems with low-grade NCOs [noncommissioned officers] and officers who remained distant from their troops.[179]

Elite forces, who constituted the bulk of the Soviet counterinsurgency troops, were generally of much higher quality than the regular forces:

> The counterinsurgency forces are carefully selected on the basis of criteria such as a clean political record, athletic ability, psychological stability. . . . Most of the counterinsurgency troops appear to be of Russian or Slavic background, and many are volunteers.[180]

BOS Insights. This case yields BOS insights in the areas of maneuver, fire support, mobility and survivability, and combat service support.

Maneuver. In Afghanistan, AFVs with lighter armor proved more useful than tanks and vehicles with heavier armor. The key capability that initially enabled the Soviets to range across the large operational area was, however, the helicopter. When the Mujahideen began receiving SA-7 and Stinger MANPADS, this Soviet advantage was significantly degraded.

Fire Support. The Soviets used virtually every fire-support system in their arsenal, short of nuclear weapons, in Afghanistan. Given the Soviet decision to refrain from massively reinforcing their troops in Afghanistan, firepower, and particularly airpower, became a force substitute:

> Soviet strategy in Afghanistan gradually came to rely almost exclusively on airpower, staking everything on airpower's capa-

---

[179] Cordesman and Wagner, *The Afghan and Falklands Conflicts*, pp. 135–136.

[180] Alexiev, *Inside the Soviet Army in Afghanistan*, p. iv.

> bilities to deliver ordnance, interdict supplies and reserves, isolate the battlefield for the rear, destroy the agricultural basis . . . and rapidly move troops from point to point.[181]

The effectiveness of Soviet airpower in Afghanistan declined after 1986, when enemy man-portable surface-to-air missiles (SAMs) began to force Soviet aircraft into higher altitudes.[182]

Mobility/Countermobility/Survivability. Because the Mujahideen were able to interdict ground LOCs, the Soviets relied on armed convoys. The theater's large distances, however, made airlift particularly important: "Air resupply was essential for rapid troop movement, reinforcement of threatened cities, response to Mujahideen attacks on roads, and a host of other problems no land-based supply system could deal with."[183]

Combat Service Support. As previously noted, the vast distances within Afghanistan created resupply difficulties. In addition, the Soviets failed to establish a logistical infrastructure at the beginning of the war to support their operations, since they expected a short war. Over time, as it became apparent that the occupation was not going to be temporary, the Soviets built the necessary logistical support infrastructure. The convoy system remained vulnerable throughout the war, however. Ground convoys often drove along difficult roads, and were always subject to interdiction by the Mujahideen. It routinely took two weeks for a ground convoy to negotiate the round trip between the Soviet border and Kabul.[184]

Characteristics of Transformation Insights. This case yields transformation insights in the areas of responsiveness, deployability, lethality, survivability, and sustainability.

Responsiveness. The Soviet forces that initially took down the Afghan government were highly capable, and their effectiveness completely surprised the Afghanis. In the aftermath of the debacle that

---

[181] Westermann, "The Limits of Soviet Airpower," p. 66.

[182] Westermann, "The Limits of Soviet Airpower," p. 76.

[183] Cordesman and Wagner, *The Afghan and Falklands Conflicts*, p. 213.

[184] Cordesman and Wagner, *The Afghan and Falklands Conflicts*, pp. 208–213.

became the Soviet war in Afghanistan, it is easy to forget how effective these forces were in the initial stage of the war. Their early success proved irrelevant in the long term, however.

Deployability. The medium-armored vehicles in the Soviet airborne forces proved highly deployable. They were airlifted into Afghanistan at the beginning of the coup de main and were able to rapidly move to critical targets. Because the BMD was transportable via helicopter, it was very useful in air-mobile operations.

Lethality. The lethality of Soviet medium-armored vehicles initially employed in Afghanistan was inappropriate for the environment. Large-caliber guns designed for conventional conflict were less useful in a counterinsurgency environment where the target was generally elusive infantry. As previously noted, the Soviets responded to the different environment by equipping medium-armored vehicles with hyperelevating, rapid-firing weapons capable of suppressing infantry targets.

Survivability. Soviet armored vehicles were vulnerable throughout the war to mines, RPGs, and heavy machine guns.

Sustainability. As noted in the section on CSS, the Soviet logistical system was severely strained by the protracted conflict.

Complex Terrain Insights. The war in Afghanistan offers insights in medium-armor performance in MOUT and mountain warfare.

Urban Terrain. Medium-armored forces made a key contribution to the ability to rapidly secure Afghan cities at the beginning of the war. They also provided protected mobility during postinvasion garrisoning operations.

Mountain Fighting. Early in the conflict, medium armor proved its usefulness in mountain warfare. Tanks and heavier-armored vehicles could not negotiate the narrow mountain trails and were mechanically unreliable in mountainous conditions. The lethality of the medium-armored vehicles had to be improved for mountain fighting by replacing existing weapons with hyperelevating, rapid-firing cannons, machine guns, and grenade launchers. The innovation of the *bronegruppa*, coupled with infantry trained for dismounted operations, further enhanced the ability of medium-armored forces to operate in mountainous terrain.

U.S. Forces in Operation Just Cause, Panama (1989)

DOTMLPF Insights. This case offers insights in the areas of organization and materiel.

Organization. The M551 Sheridan company from the 3rd Battalion, 73rd Armor, was deployed in sections (of two vehicles each), rather than as platoons, with Task Force Bayonet and Task Force Pacific. CALL noted that this concept of employment

> allowed the simultaneous engagement of many of the D-Day targets with the infantry supported by the shock effect and fire power of a mobile protected gun system. The shock effect and firepower were critical in military operations on urbanized terrain (MOUT) fighting, at roadblocks, for fixed site security and convoy escort.[185]

Materiel. M113 APCs and M551 Sheridans, both relatively old vehicles, performed well during Operation Just Cause. M113s proved an "effective gun platform, armored carrier, evacuation and roadblock vehicle. The cargo hatch in the rear allowed troops with body armor to conduct 360-degree surveillance and engage snipers on rooftops."[186] M551 Sheridans provided direct fire support and, using their 152-mm guns, blew holes in buildings for infantry assaults. Both the M113 and the M551 proved useful in intimidating PDF forces, establishing roadblocks, and providing "fire-power demonstrations."[187]

BOS Insights. This case yields BOS insights in the areas of maneuver, fire support, and mobility and survivability.

Maneuver. Armored vehicles provided protected mobility for the forces engaged in Operation Just Cause. The LAVs in Task Force Semper Fi enabled the task force to rapidly respond to the mission of reducing the PDF in La Chorrera. Medium-armored vehicles were able

---

[185] Center for Army Lessons Learned, *Operation Just Cause Lessons Learned*: Vol. II, *Operations* (No. 90-9), Fort Leavenworth, Kan.: U.S. Army Combined Arms Command, 1990, p. II-2.

[186] Center for Army Lessons Learned, *Operation Just Cause Lessons Learned*: Vol. III, p. III-15.

[187] CALL, *Operation Just Cause Lessons Learned*: Vol. III, p. III-15.

to traverse Panamanian bridges that could not have supported U.S. heavy armor.[188]

Fire Support. The weapons provided by the M113s, M551s, and LAV-25s were a critical source of fire support throughout Operation Just Cause. The fires provided by attack helicopters and AC-130 gunships minimized collateral damage in urban areas. On occasion, M551 Sheridans provided critical fires when aerial platforms were not able to engage targets obscured by dense foliage.[189]

Mobility/Countermobility/Survivability. U.S. medium-armored vehicles were very effective both in establishing roadblocks to impede enemy movement and in reducing roadblocks established by the enemy.

Characteristics of Transformation Insights. This case yields transformation insights in the areas of responsiveness, deployability, lethality, and survivability.

Responsiveness. U.S. medium-armored forces, which had been largely prepositioned in Panama, participated in a highly integrated operation to take down Noriega's government. M551 Sheridans were dropped into areas where their fires often provided critical capabilities. However, the majority of the dispersed operations outside of Panama City were initially executed by light airborne and ranger forces.

Deployability. Many of the armored vehicles used during Operation Just Cause had arrived in Panama prior to D-Day. Nevertheless, the deployability of the M551 Sheridans, which were airdropped with Task Force Pacific, enabled them to make significant contributions to the task force. The M551 Sheridans were very effective as mobile guns, and were of particular use in heavily forested areas that precluded the use of aerial platforms.

Lethality. The weapons on the M113s, M551s, and LAV-25s proved highly effective in supporting Operation Just Cause. Given the restricted rules of engagement, the ability of these platforms to aim precise direct fire against many categories of targets was a critical enabler, particularly for dismounted infantry operations.

---

[188] Phillips, *Operation Just Cause: The Incursion into Panama*, p. 45.

[189] Sherman, "Operation Just Cause: The Armor-Infantry Team in the Close Fight," p. 35.

Survivability. M113s, M551s, and LAVs proved very survivable against the capabilities of the PDF. As previously noted, the M113 provided protected mobility for infantrymen was particularly useful in MOUT. Finally, soldiers improved the survivability of the M113s by piling layers of sandbags on top of them. Soldiers also put concertina wire on M113s to protect them from tampering and to make them more useful in roadblocks.[190]

Complex Terrain Insights. Armored platforms were used in MOUT and jungle terrain during Operation Just Cause.

Urban Terrain. As previously noted, the medium-armored platforms used during Operation Just Cause provided protected maneuver and fire support to infantry forces engaged in MOUT. The 152-mm gun of the M551 Sheridan was particularly useful in blowing holes in walls to enable infantry maneuver, and in reducing obstacles.

Jungle. M551 Sheridans provided fire support to Task Force Pacific in the heavily forested areas of its AO. This was important because dense foliage occasionally made it impossible for AC-130 gunships to provide fire support.

Russia in Chechnya (1994–2001)

DOTMLPF Insights Chechnya I. This case offers insights in the areas of doctrine, training, leader development, organization, materiel, and personnel.

Doctrine. The Russian Army doctrine for MOUT was largely inappropriate for the situation in which it found itself during the initial battle for Grozny. The doctrine

> presented two options: if a city was defended, it was to be bypassed; if it was not, it could be taken from the march. In the latter case, entering troop formations would conduct a show of force rather than fight. Tanks would lead followed by mounted and dismounted infantry.[191]

---

[190] Sherman, "Operation Just Cause: The Armor-Infantry Team in the Close Fight," p. 35.

[191] Oliker, *Russia's Chechen Wars*, p. 5.

This doctrine clearly failed in Grozny, and new tactics had to be developed during combat. These new tactics are discussed under the section on materiel, below.

Doctrine for mountain fighting was largely a continuation of the Soviet doctrine developed for Afghanistan. It is discussed in the Chechnya II section on doctrine, below.

Training. Absent appropriate doctrine, the Russian Army was not trained for the conditions it encountered in Chechnya. This lack of training was exacerbated by the pitiable state of the Russian Army when the war began:

> The Russian Army had been operating with little money and bare bones logistical support. It had not conducted a regiment- or division-scale field training exercise in over two years, and its battalions were lucky to conduct field training once a year. Most battalions were manned at 55% or less.[192]

While the state of training was abysmal in general, it was virtually nonexistent for MOUT:

> One Russian officer noted that a rehearsal for taking a built-up area had not been conducted in the last 20 to 25 years, which contributed to decisions such as sending the force into the city in a column instead of in combat formation.[193]

At the level of the individual soldier, the Russian force that entered Chechnya was similarly bedraggled:

> The available infantry had been thrown together, and many did not know even the last names of their fellow soldiers. They were told that they were part of a police action. Some did not have weapons. Many were sleeping in the carriers even as the columns rolled into Grozny. Tank crews had no machine gun ammunition.[194]

---

[192] Grau, "Russian Urban Tactics."

[193] Thomas, "The Battle of Grozny."

[194] Thomas, "The Battle of Grozny."

Leader Development. Many Russian junior officers and noncommissioned officers who initially deployed to Chechnya were generally inexperienced and poorly trained.[195] This situation improved somewhat with the introduction of elite units.

Organization. As previously noted, the Russian units that comprised the storm detachments that entered Grozny were thrown together at the last moment. There was little unit cohesion. The invading Russian Army was "a rag-tag collection of various units, without an adequate support base."[196] Furthermore, the Russian Army

> was forced to combine small units and send them to fight. Infantry fighting vehicles went to war with their crews, but with little or no infantry on board. In some cases, officers drove because soldiers were not available.[197]

This situation also improved with the introduction of more-cohesive elite units.

Materiel. Russian armored vehicles, designed for a war with NATO, operated with difficulty in the conditions found in Chechnya. Tanks and other armored vehicles, which had been designed to fight NATO tanks and armored vehicles, were armored most heavily in front. Thus, they were vulnerable to side, top, and rear attack, and Chechen rebels exploited these vulnerabilities to great effect, using a tactic where "five or six hunter-killer teams simultaneously attacked a single vehicle."[198] The rebels also "dropped bottles of jellied gasoline on top of vehicles."[199] Even tanks were destroyed by rebel RPGs.[200] The

---

[195] Speyer, "The Two Sides of Grozny," p. 73.

[196] Speyer, "The Two Sides of Grozny," p. 73.

[197] Speyer, "The Two Sides of Grozny," p. 73.

[198] Lester M. Grau, "A Weapon for all Seasons: The Old But Effective RPG-7 Promises to Haunt the Battlefields of Tomorrow," Fort Leavenworth, Kan.: Foreign Military Studies Office, 1998.

[199] Lester M. Grau and Jacob W. Kipp, "Urban Combat: Confronting the Specter," *Military Review*, Vol. 79, No. 4, July–August 1999, p. 15.

[200] Grau and Kipp, "Urban Combat," p. 15; Grau, "A Weapon for all Seasons." Grau notes the following:

elevation and depression of many Russian armored-system weapons were "incapable of dealing with hunter-killer teams fighting from basements and second or third-story positions." [201] This was particularly a problem with tank main guns and coaxially mounted machine guns. Flexibly mounted machine guns were more useful, but they exposed gunners to sniper fire.

The Russians adapted to the realities of combat in the urban canyons of Grozny. They modified their tactics, choosing to lead armor with infantry, use artillery, and even employ self-propelled howitzers and BM-21 multiple rocket launchers as direct-fire weapons against rebel strongpoints.[202] They also tried to reduce the vulnerability of tanks to top, side, and rear RPG attacks by installing reactive armor and wire mesh cages (not unlike the U.S. Army's "RPG screens" in Vietnam) on vehicles. Furthermore, the Russians began employing ZSU-23-4 and 2S6 antiaircraft systems, which were able to elevate and depress to attack rebels in multistory buildings, to compensate for the shortcomings in their other weapons. These air defense vehicles were lightly armored and vulnerable, however, and the Chechens learned to attack them first.[203] Additionally, the Russians "found that wheeled armored personnel carriers (BTRs) were often better suited for urban combat than tracked armored personnel carriers (BMPs)."[204] Finally, the Russians used BTRs to resupply forces engaged in Grozny because unarmored trucks were highly vulnerable to rebel fire.[205]

> During the first month of the conflict, Russian forces wrote off 225 armored vehicles as nonrepairable battle losses. . . . 10.23% of the armored vehicles initially committed to the campaign. The bulk of these losses were due to shoulder-fired antitank weapons and antitank grenades.

[201] Lester M. Grau, "Russian-Manufactured Armored Vehicle Vulnerability In Urban Combat: The Chechnya Experience," *Red Thrust Star*, January 1997.

[202] Grau and Thomas, "Russian Lessons Learned." The Chechens countered this new tactic by "hugging" Russian forces to get them to cease air and artillery support for fear of fratricide.

[203] Grau and Kipp, "Urban Combat," p. 15. Grau; "Russian-Manufactured Armored Vehicle Vulnerability."

[204] Grau and Kipp, "Urban Combat," p. 15.

[205] Grau and Kipp, "Urban Combat," p. 15.

The Russians experienced difficulties with their communications equipment in Grozny. The principal cause was

> simply the vertical obstacles posed by urban structures. High-rise buildings and towers impeded transmissions, especially those in the high to ultra high frequencies. Communications officers had to consider the nature of radio wave propagation and carefully select operating and alternate frequencies, and they had to consider the interference caused by power transmission lines, communications lines, and electric transportation contract systems.[206]

The Chechens, on the other hand, used commercial off-the-shelf equipment (e.g., Motorola radios and Iridium satellite systems) to great effect.[207]

Personnel. The quality of the Russian soldier was generally quite low: "Approximately 85% of Russian youth were exempt or deferred from the draft, forcing the army to accept conscripts with criminal records, health problems or mental incapacity."[208] Morale was low, and over two-thirds of the infantry conscripts had less than six months military experience.[209] As one author noted,

> the strength of the Russian Army was material and its weakness was human. It had inherited from the Soviet Army an arsenal of modern, if no longer state-of-the-art, weapons but its soldiers were poorly trained and badly motivated.[210]

Indeed, there were cases in Grozny when "Russian conscript Infantry simply refused to dismount and often died in their BMP without ever firing a shot."[211]

---

[206]Thomas, "The Battle of Grozny."

[207]Grau and Thomas, "Russian Lessons Learned."

[208]Grau, "Russian Urban Tactics."

[209]Speyer, "The Two Sides of Grozny," p. 73.

[210]Michael Orr, "Better or Just Not So Bad? An Evaluation of Russian Combat Effectiveness in the Second Chechen War," in Aldis, *The Second Chechen War*, p. 92.

[211]Grau and Thomas, "Russian Lessons Learned."

DOTMLPF Insights Chechnya II. This case offers insights in the areas of doctrine, training, organization, materiel, and personnel.

Doctrine. Doctrine was the source of the most crucial Russian error in this conflict. The belief that artillery barrages and air strikes would be sufficient to enable the Russians to avoid an urban fight proved to be a costly mistake. It led the Russians to fail to train their forces for urban combat, and led them to develop an approach to the city that relied predominantly on police forces and procedures, rather than on the close combat they ultimately faced.

Russian doctrine for fighting in the mountains of Chechnya was based on experiences in Afghanistan, Tajikistan, and in Chechnya five years before. The Russians knew that the key was to control the LOCs, mountain passes, and commanding heights. This often meant employing paratroopers and light, small-unit operations, as well as relying on rotary-wing aviation for a wide range of roles.[212] The Russians also maintained the belief in overwhelming firepower so evident in their approach to urban areas. Massive artillery and air strikes on areas where rebels were believed to be based preceded the delivery of Russian forces to the area.[213] Armor served numerous purposes, including attempts to seal off roads, create roadblocks, and serve as a form of artillery. However, as armor's reach was limited by terrain, infantry troops often had little armor support. A key infantry weapon was the *Shmel* flamethrower, which had proven effective against tunnels and caves in Afghanistan and was used in a similar manner in Chechnya.[214]

---

[212]Aleksandr Romanyuk, "Zakhvat Perevala [Capturing a Mountain Pass]," *Aremeiskii Sbornik*, December 2000, pp. 37–39.

[213]Dmitri Nikolaev, "Spetzoperatziya Zavershilas', a Voyna Prodolzhayetsya [Special Operation Complete, But War Continues]," *Nezavisimoye Voyennoye Obozreniye*, March 3, 2000.

[214]Dmitri Nikolaev, "Voyska Idut V Gori [Troops Going to Mountains]," *Nezavisimoye Voyennoye Obozreniye*, February 11, 2000. On the use of the *Shmel* flamethrower in Afghanistan, see Lester W. Grau and Ali Ahmad Jalali, "Underground Combat, Stereophonic Blasting, Tunnel Rats and the Soviet-Afghan War," *Engineer*, Vol. 28, No. 4, November 1998, pp. 20–23. See also Oliker, *Russia's Chechen* Wars, p. xxiii, which describes the *Shmel* as a

> [n]ew generation 'flamethrower.' 11-kg., single-shot, disposable, 600-meter range weapon carried in packs of two by ground forces. The warhead is equipped with a 'ther-

Training. The Russians made a conscious decision not to train their forces for urban combat. Of the many late 1990s Russian exercises that appeared to focus on operations in the Caucasus, none included a sizable urban component. Although forces prepared for mountain fighting and counterterrorist operations, they did not train for close combat in an urban environment. Once fighting was underway, however, the Russians established training centers on the outskirts of towns where they were based, mirroring a practice from Chechnya I. In spite of initial promises to send only seasoned troops into combat in Grozny, there were simply not enough veterans to go around, and it soon became clear that raw recruits were being sent to fight in the city's streets and buildings.

The Russians were better prepared for mountain fighting. Russian preparation between the wars had focused significantly on mountain combat and had included a number of training exercises that incorporated a mix of forces. Specific units (particularly the paratroopers, but also the naval infantry and some motor rifle troops) received specialized training for mountain fighting, and the border guards and other reinforcements sent to Chechnya in 2000 also received some additional training before being sent into battle. Reportedly, a unit was preparing for combat at the Dar'yalsk mountain training center in North Ossetia during the Grozny fighting and was dispatched to the mountains in February.[215] However, the increased training was not universal and, according to some reports, was often insufficient.[216]

Organization. As previously noted, force coordination was consistently better in 1999–2000 than in 1994–1995. Troubling cleavages existed, however, between the MoD and MVD, and especially between the MoD and the various militias, including a loyalist militia led by a former Grozny mayor. There were reports of fratricide as a result of the

mobaric' incendiary mixture, a fuel-air explosive, which upon detonation produces an effect comparable to that of a 152mm artillery round.

[215] Nikolaev, "Forces Heading to the Mountains"; Romanyuk, "Zakhvat Perevala [Capturing a Mountain Pass]."

[216] Vadim Udmantzev, "Polkoviye 'Zagogulini' [Regimental Bumbling]," *Nezavisimoye Voyennoye Obozreniye*, Internet edition, February 2, 2001.

Russian forces' occasional inability to distinguish loyalists from the enemy.

In their organization for combat, the Russians had learned lessons from both their own World War II experience and from their enemy. They set up attack ("storm") groups of 30 to 50 men and broke these groups into even smaller teams of a handful of men each. These smaller teams might include soldiers armed with an RPG, an automatic rifle, and a sniper rifle, and include two additional men armed with automatic weapons. Other storm group components included soldiers armed with Shmel flamethrowers, artillery and aviation forward-observers, sappers, and reconnaissance personnel.

Throughout 2000, the command structure placed the MoD at the head of a multiagency-based force; in 2001, command was transferred to the FSB. In the mountains, the force included air and air-defense forces, border troops, ground forces (including motor rifle troops), army aviation, naval infantry forces, special operations forces, MVD units and police, some FSB personnel, and paratroopers (a separate service in the Russian military). With the exception of the MVD troops, all of these units ordinarily reported to the MoD. Coordination difficulties surfaced and persisted, particularly between air and ground units (aviators complained about spotter incompetence) and between the MoD and the MVD.[217]

Materiel. The better use of armor in MOUT—achieved through the decision not to lead with armor and through armor's close cooperation with infantry—improved its survivability. The Russians admitted to the loss of only one tank in Grozny in 1999–2000 fighting. That said, an unknown number of other Russian armored vehicles were destroyed. Rebel tactics remained largely the same compared to

[217]Oliker, *Russia's Chechen Wars*, pp. 51, 57; "Na Argunsoye Ushchel'ye Sbrasivayut Ob'yemno-Detoniruyushchiye Bombi [Fuel-Air Bombs Being Dropped on Argun Region]"; Vladimir Matyash, "Uchast' Banditov V Gorakh Predpreshena [Fate of Bandits in Mountains Predetermined]," *Kraznaya Zvezda*, February 9, 2000; "Osnovniye Boyi V Chechenskikh Gorakh Razvernutsya Cherez Dva Dnya [Main Battles in Chechen Mountains To Begin in Two Days]"; "V Gorniye Rayoni Chechni Perebrosheno Podkrepleniye [Reinforcements Sent to Mountainous Regions of Chechnya]"; Sergey Val'chenko and Konstantin Yur'yev, eds., "Krugliy Stol AS: Goryachiy Vozdukh Kavkaza [AS (Armeiskii Sbornik) Roundtable: Hot Air of the Caucasus]," *Armeiskii Sbornik*, February 2001, pp. 24–32.

Chechnya I, and medium-armored vehicles and the personnel within them once again proved vulnerable to small arms, RPGs, artillery, and mines. As in the first war, soldiers jury-rigged protective wire cages, sandbags, boxes, and other defenses.[218]

In the mountains, Russian tanks were vulnerable to mines, various explosives, and the enemy's portable antitank weapons. The Russians' laborious and time-consuming approach to mine-clearing slowed troop movement. IFVs and BTRs were vulnerable to "almost any artillery," and to mines and antitank grenades. Armored cars and trucks were poorly protected against bullets and shell fragments. As in the 1994–1996 war, poor roads and difficult conditions meant that even the best wheeled trucks and automobiles (e.g., Zil-131s, Gaz-66s, and UAZ-452s) were unreliable. More "civilian-type" vehicles, such as the Maz-500, Zil-130 and Gaz-53, experienced even greater difficulty navigating the terrain, and suffered high rates of frame and carriage breakage. The Ural 4320 (a 6X6 off-road, all-terrain cargo truck, which also comes in a 4X4 variant), however, reportedly did quite well. In both Chechen wars, the BTR-60PB and BTR-70 were found to be unreliable. Evacuating heavy armor was problematic.[219]

Personnel. Stories of hazing, drug abuse, and selling weapons to the enemy for money or narcotics were prevalent, though perhaps slightly reduced compared to Chechnya I levels. Although better and more reliable pay and much higher military support for the mission improved morale in the early stages of Chechnya II, this improvement was tempered as casualties mounted and the fight dragged on.

While some of the forces deployed to Chechnya were no doubt well-trained professionals, others, particularly among both the enlisted and the junior officers of the motor rifle troops, had little experience and only minimal training. Insufficient levels of training for tank and BMP drivers and mechanics likely had an impact on the poor performance of these vehicles. Soldiers and technicians were not trained in key spe-

---

[218] Andrei Mikhailov, "They've Learned To Use Tanks," *Nezavisimaya Gazeta*, May 25, 2000.

[219] Mikhailov, "They've Learned To Use Tanks."

cialties, including communications.[220] One journalist's account of the mishaps, injuries, and deaths among one Russian regiment (the 291st Motor Rifle Regiment of the 42nd Motor Rifle Division) in Chechnya in April, May, and June of 2000 revealed that most of these casualties resulted from poor equipment handling, internal scuffles between soldiers, and other noncombat causes. Moreover, this lack of professionalism was evident in professional soldiers as well as conscripts. The same journalist reported high rates of alcohol abuse and noted that the statistics for other units of the 42nd Division were not dissimilar.[221]

BOS Insights Chechnya I. This case yields BOS insights in the areas of maneuver, fire support, C2, intelligence, and mobility and survivability.

Maneuver. The conditions in Grozny and other Chechen urban centers severely restricted armored vehicle maneuver, as did fighting in the mountains.

Fire Support. After the initial debacle in Grozny, the Russians came to rely on the use of heavy firepower. Artillery was "used to compensate for poor infantry performance."[222] Additionally,

> the Russians began to use massed artillery routinely as a substitute for maneuver combat. Previous Russian concerns about civilian casualties vanished in the face of the limited success from mass artillery strikes against the Chechens.[223]

Following the capture of Grozny, the Russians developed a firepower-intensive tactical method for dealing with rebel villages. Russian forces would surround and cut off the town. After notifying the townspeople of their intent to storm the town, they would "shell the village until

---

[220]Oliker, *Russia's Chechen* Wars, pp. 53–54; Udmantzev, "Polkoviye 'Zagogulini' [Regimental Bumbling]"; Val'chenko and Yur'yev, eds., "Krugliy Stol AS: Goryachiy Vozdukh Kavkaza [AS (Armeiskii Sbornik) Roundtable: Hot Air of the Caucasus]."

[221]Udmantzev, "Polkoviye 'Zagogulini' [Regimental Bumbling]."

[222]Speyer, "The Two Sides of Grozny," p. 75.

[223]Gregory J. Celestan, "Wounded Bear: The Ongoing Russian Military Operation in Chechnya," Fort Leavenworth, Kan.: Foreign Military Studies Office, August 1996.

return fire ceased and then move in. The Chechens would redeploy to another village and wait for the next column of Russian vehicles."[224]

As a consequence of the profligate use of firepower against an elusive enemy, the Russians turned the population against them. Because they were

> [u]nable to accurately target the Chechen rebels . . . and crush the Chechen center of gravity, Russian forces adopted a "shot gun" approach. They delivered tons of ordnance in the hope of taking out individual Chechen snipers. This unjudicious [*sic*] employment of combat power served to alienate a large percentage of the potentially neutral Chechen population and transformed them into active combatants.[225]

A Russian political commentator noted that the Russians had "not won anything in Chechnya; rather we have acted like a blindfolded, robust child, thrashing around blindly with an ax."[226]

Command and Control. In addition to the communications equipment problems previously noted, Russian C2 were complicated by the command and organizational arrangements made prior to the deployment to Chechnya. The Russian force, composed as it was of units from different agencies (MoD, MVD, etc.) that had not trained together, experienced significant coordination and unity-of-command problems.[227]

Intelligence. Russian intelligence efforts were inadequate during Chechnya I: "Simply put, the Russians did not do a proper intelligence preparation of the battlefield."[228] In the initial assault on Grozny, they

> had almost no information about the situation in the city, especially from human intelligence sources. Military intelligence did

---

[224]Celestan, "Wounded Bear."

[225]Raymond C. Finch, "Why the Russian Military Failed in Chechnya," Foreign Military Studies Office, Special Study No. 98-16, 1998.

[226]Finch, "Why the Russian Military Failed."

[227]Celestan, "Wounded Bear;" Finch, "Why the Russian Military Failed in Chechnya."

[228]Thomas, "The Battle of Grozny."

> not delineate targets for air and artillery forces, and electronic warfare resources were not used to cut off President Dudayev's communications. Reconnaissance was poorly conducted, and Chechen strong points were not uncovered. There was little effective preliminary reconnaissance of march routes, reconnaissance amounted to passive observation, and reconnaissance elements appeared poorly trained.[229]

Further complicating matters for the Russians was a severe shortage of accurate, appropriately scaled maps.[230] Indeed, "[o]nly a few large-scale maps were available, and there were no maps available to tactical commanders." Furthermore,

> [e]ssential aerial photographs were not available for planning, because Russian satellites had been turned off to save money and few aerial photography missions were flown. Lower-level troop commanders never received vital aerial photographs and large-scale maps.[231]

The Chechen rebels, on the other hand, generally had the support of the local people, who often informed them of Russian movements and activities. The rebels were also intimately familiar with the terrain—they were fighting on their home ground.[232]

Mobility/Countermobility/Survivability. The poor infrastructure in Chechnya limited Russian mobility. This was an issue for armored vehicles: "[M]any of the armored vehicle drivers had enormous difficulty driving on the thin, muddy asphalt roads which are the main highways through the region."[233]

---

[229]Thomas, "The Battle of Grozny."

[230]Speyer, "The Two Sides of Grozny," p. 78.

[231]Grau, "Russian Urban Tactics."

[232]Grau, "Russian Urban Tactics"; and Anatoliy S. Kulikov, "The Chechen Operation," in Glenn, ed., *Capital Preservation*, 2001, p. 51.

[233]Celestan, "Wounded Bear."

Combat Service Support. Support to fielded forces was abysmal, because Russian "[l]ogistics support for the operation was not developed for sustained combat operations in Chechnya."[234] Indeed,

> Russian soldiers were inadequately fed, clothed and sheltered . . . . Untrained soldiers were sent into combat without or with substandard equipment. . . . Some Russian soldiers surrendered to the enemy without a fight, or sold their arms to the Chechens for food or drugs and alcohol.[235]

BOS Insights Chechnya II. This case yields BOS insights in the areas of maneuver, fire support, C2, intelligence, and mobility and survivability.

Maneuver. With the overall emphasis on firepower and destruction that the Russians took to the second Chechen war, it is not surprising that armor was often relegated to an assault-gun role or employed as additional artillery. The smaller-unit tactics were effective for seizing territory, although the Russians were to learn once again that in the urban environment, the front line is an amorphous concept, and frequent attacks from what were believed to be secure areas made it difficult to hold territory.

Rotary-wing aviation was the key component of operations in the mountains. Helicopters provided close air support, delivered troops (including motor rifle troops) to the mountains, evacuated dead and wounded, delivered supplies, and carried out other tasks. While various armored and unarmored automobiles, trucks, BTRs, BMPs, IFVs, tanks, and other vehicles were used where possible, they were significantly limited by the terrain and, as previously noted, were vulnerable to a range of enemy weaponry. However, helicopters also operated within constraints. At higher altitudes, for instance, they could carry less weight. There were relatively few places for them to land safely. Enemy air defenses affected how high they could fly. Helicopters were assigned to air tactical groups composed of two to four Mi-24 attack

---

[234] Celestan, "Wounded Bear."

[235] Finch, "Why the Russian Military Failed."

helicopters and one or two Mi-8 transport helicopters. The groups reported to ground-force commanders and, in theory, were coordinated by controllers on the ground. (Helicopter crews complained about the inexperience of the controllers, and about the fact that there were not enough of them.) Supporting MVD units was particularly difficult because the units were inexperienced with aviation and often operated under incompatible communications protocols. Even many ground force commanders had little knowledge of how to work effectively with aviation.[236]

Fire Support. Massive fire support was the norm for Russian operations in the second Chechen war, and it caused very high levels of collateral damage. The Russians believed that increased firepower was correlated with decreased casualties among Russian forces, and behaved accordingly:

> Artillery . . . was the basis of Russian combat in both Grozny and Chechnya as a whole in 1999–2000. Artillery was the day and night, all-weather tool for keeping the enemy at a distance and, it was hoped, for protecting Russian soldiers from close combat. Encircled towns were shelled into submission, artillery "prepared" parts of a city or town for ground force entry, and soldiers felt comfortable calling for it whenever they met with resistance.[237]

In the mountains, while Russian forces continued to believe that overwhelming firepower was an effective solution to many problems, poor visibility, problems with air-ground coordination, and poor communication links sometimes hampered the ability of air and artillery assets to effectively support troops. In principle, air and artillery strikes were supposed to wipe out resistance before ground troops entered an area. In practice, although a great many bombs were dropped and much artillery was used, ground forces were often unsure of the conditions they were heading into. There were reports of fixed-wing aviation

---

[236]Oliker, *Russia's Chechen Wars*, pp. 51, 56, 57; Romanyuk, "Zakhvat Perevala [Capturing a Mountain Pass]," pp. 37–38; Val'chenko and Yur'yev, eds., "Krugliy Stol AS: Goryachiy Vozdukh Kavkaza [AS (Armeiskii Sbornik) Roundtable: Hot Air of the Caucasus]."

[237]Oliker, *Russia's Chechen Wars*, pp. 57–58.

declaring targets destroyed and forces moving into the area to discover that the targets remained active. Ground controllers often failed to communicate target locations effectively due to poor communications and inexperience. These problems may have resulted in fratricide.[238]

Command and Control. As previously discussed, C2, though much improved compared to the first war, exhibited continued problems in the joint operations of the very disparate Russian forces. One of the great success stories, however, was the vastly improved coordination between air and ground forces (although both ground and air units continued to report problems). Communications remained challenging at times due to incompatible protocols, trouble with air-ground connections, and a continuing failure by Russian forces to communicate securely, which allowed the enemy to learn of Russian plans.[239]

Intelligence. Russian intelligence failed to accurately assess the size of the remaining enemy resistance in Grozny. As they did in 1994, Russian forces simply assumed the best and acted accordingly.

Accurate intelligence proved difficult to obtain outside the cities. Helicopter pilots did not appreciate being sent out on reconnaissance duty into unknown terrain, and tracking enemies in their home mountains and passes was often impossible. Of course, this made delivering forces and carrying out air and ground attacks that much more difficult. The Russians probably intercepted enemy communications on occasion, but the rebels seemed to intercept Russian communications far more frequently, or at least more effectively, particularly because Russian forces often failed to communicate securely.[240]

---

[238]Oliker, *Russia's Chechen Wars*, pp. 51, 56, 57; Val'chenko and Yur'yev, eds., "Krugliy Stol AS: Goryachiy Vozdukh Kavkaza [AS (Armeiskii Sbornik) Roundtable: Hot Air of the Caucasus]."

[239]Oliker, *Russia's Chechen Wars*, pp. 51–52. See also, Val'chenko and Yur'yev, eds., "Krugliy Stol AS: Goryachiy Vozdukh Kavkaza [AS (Armeiskii Sbornik) Roundtable: Hot Air of the Caucasus]."

[240]Oliker, *Russia's Chechen Wars*, pp. 51–52; Val'chenko and Yur'yev, eds., "Krugliy Stol AS: Goryachiy Vozdukh Kavkaza [AS (Armeiskii Sbornik) Roundtable: Hot Air of the Caucasus]"; A. Aseev and A. Malyugin, "Na Tryokh Kitakh [On Three Whales]," *Armeiskii Sbornik*, November 2000, pp. 33–35.

Combat Service Support. CSS had improved since the first Chechen war. Although many units reported insufficient supplies and food, their plight paled in comparison to the tales of Russian soldiers starving in Chechnya in the mid-1990s. To accomplish Chechen II levels of support, however, the Russians stretched supply lines and reserves tremendously.[241]

Support in the mountains was the most problematic. Some special operations forces reported waiting up to a week for supplies as helicopter units waited for the fog to lift.[242] Soldiers often found that what they received bore little resemblance to what guidelines called for them to have.[243] Fuel shortages occurred, and commanders limited how often soldiers could use their vehicles. Spare parts were also in short supply, and vehicles were often cannibalized.[244]

Russian supplies frequently ran low, and there is no doubt that soldiers supplemented their stores by stealing from the local populace (which they might have done out of greed and malice as much as out of need).[245] Fuel shortages, problematic as early as the Dagestan campaign, only got worse.[246]

Characteristics of Transformation Insights. The Chechnya I and Chechnya II cases yield similar transformation insights in the areas of lethality and survivability.

Lethality. As previously noted, Russian armored vehicles had difficulty engaging targets in multistory buildings.

Survivability. Russian armored vehicles, designed as they were for a NATO environment, proved vulnerable to tank hunter-killer teams. These teams attacked the relatively lightly armored tops, sides, and rears of Russian armored vehicles.

---

[241] Oliker, *Russia's Chechen Wars*, pp. 61–62.

[242] Boikov, "Luchshe Gor Mogut Bit'… [Better Than Mountains Could Be…]."

[243] Val'chenko and Yur'yev, eds., "Krugliy Stol AS: Goryachiy Vozdukh Kavkaza [AS (Armeiskii Sbornik) Roundtable: Hot Air of the Caucasus]."

[244] Udmantzev, "Polkoviye 'Zagogulini' [Regimental Bumbling]."

[245] Oliker, *Russia's Chechen Wars*, pp. 61–62.

[246] Oliker, *Russia's Chechen Wars*, p. 54.

Complex Terrain Insights. The Chechnya I and Chechnya II cases yield several insights in the area of complex terrain performance.

Urban Terrain in Chechnya I. As previously noted, the principal MOUT fight in the first Chechen war involved a protracted assault on Grozny. Upon entering this battle, the Russians unreasonably expected that Grozny would fall rapidly in a coup de main, much like Prague in 1968 or Kabul in 1979. In essence, they expected the Chechens to cower in the face of their armored presence. This did not happen. Instead, tenacious Chechen rebels waged a skillful MOUT fight against poorly trained Russian troops.

Russian tanks and medium-armored vehicles had difficulty engaging rebels located in upper stories and basements—their weapons would not elevate or depress sufficiently to engage targets. Russian antiaircraft artillery vehicles (i.e., ZSU-23-4s and 2S6s) were effective in engaging targets located in upper-story positions, but were vulnerable to enemy fire because of their thin armor. All Russian armored vehicles were vulnerable to attacks by Chechen hunter-killer teams that were armed with RPG-7 or RPG-18 shoulder-fired antitank rocket launchers. The Chechen tactic of choice was "to trap vehicle columns in city streets where destruction of the first and last vehicles will trap the column and allow its total destruction."[247] Following their disastrous initial assault on Grozny, the Russians modified their tactics, as previously noted.

Urban Terrain in Chechnya II. Many of the problems the Russians faced in Grozny had less to do with their force mix than with their poor preparations and faulty assumptions. They believed that artillery and air strikes could decimate the enemy such that urban combat would be unnecessary, and they structured and trained the force sent into Grozny accordingly. The force was heavy on MVD and local militia units and the conscript forces of the motor rifle troops. It was light on the capable snipers and personnel who were experienced in the small-unit tactics that urban combat calls for. Moreover, Russian underestimation of enemy force size and capacity led to the deployment of too small a force initially, and to the need for large numbers of reinforcements.

---

[247] Grau, "Russian-Manufactured Armored Vehicle Vulnerability."

Armor was better-utilized in Chechnya II because of improvements to its integration with infantry and other arms. Although it once again exhibited vulnerabilities to mines, RPGs, explosives, small arms, and other weapons, better tactics and a repeat of the jury-rigging of vehicle defenses that soldiers undertook in 1994–1996 resulted in less frequent destruction of Russian armored vehicles, although levels were still too high. The improved approach to armor did not appear to carry over into some later urban battles, including the one in Komsomolskoye in 2000, where armor losses were more significant.

Finally, the Russian belief that massive firepower would save their soldiers' lives resulted in tremendous collateral damage and no doubt numerous civilian casualties. Moreover, Russian forces continued to sustain casualties, particularly from snipers, which each side relied on heavily as the battle ground on.[248]

Mountain Fighting in Chechnya I. After the fall of Grozny, the Russian Army fought to control other major Chechen cities and towns; they largely accomplished this goal by May 1995. The rebels moved into the mountains where they continued a guerrilla-style war against the Russians. As previously noted, the Russian approach to mountain towns that harbored rebels was to shell the villages until the rebels pulled out, then take the village. Meanwhile, rebel forces would move on to another village, and the scenario would be repeated.

A friendly local populace and difficult terrain sustained the rebel resistance in the mountains. Nevertheless, rebel forces were feeling pressure from the Russian forces, and were on the run. They were also losing fighters to desertion. This situation changed when rebel forces began making forays into Russia itself. In June 1995 they attacked Budennovsk, a town some 70 km north of Chechnya. In the aftermath of this raid, rebel morale and recruitment improved. Still, the period from May 1995 until the end of the first Chechen war was largely characterized by inconclusive Russian operations and low-level rebel

---

[248]Oliker, *Russia's Chechen Wars*, p. 48. See also Kulikov, "The Chechen Operation," in Glenn, ed., *Capital Preservation*, 2001, p. 57.

activity, with the occasional major Russian foray against high-visibility targets, like Grozny.[249]

Mountain Fighting in Chechnya II. The Russian military leadership took great pains to emphasize in public announcements that fighting in the mountains would involve well-trained, specialized troops. However, the motor rifle units that constituted the bulk of the Russian conscript-based military force were clearly involved in the fighting (along with MVD troops), particularly in the urban fighting in mountain towns.

The Russian force in the mountains relied heavily on rotary-wing aviation for a variety of tasks, including supply, firepower, and evacuating the wounded. Fixed-wing aircraft also played an important role, and carried out many bombing missions. Although armored trucks and automobiles, BTRs, APCs, and some tanks were present, their utility was constrained by the complex terrain.

The rebel approach relied on heavy mining, particularly of approaches to populated areas and roads (control of which was a primary Russian objective). They carried out raids on border troops deployed in the region and were most effective at inflicting casualties when they were able to ambush Russian forces that were on the move (especially in road-bound columns) or preparing strongpoints. Rebel personnel were armed with RPGs, antitank missiles, mortars, small arms, and, according to some reports, flamethrowers.[250] Their air defense weapons included various SAMs, *Shilka* air-defense guns, and other mobile systems, although reports vary on just how many of these they had. Heavy-caliber machine guns and antitank missiles were also used for air defense. Rebels usually targeted the back of a helicopter as it turned so that the pilot could not identify the source of fire. Whenever possible, they targeted the sides, back, and top of aircraft.

---

[249]Oliker, *Russia's Chechen Wars*, pp. 28–31; Celestan, "Wounded Bear;" Kulikov, "The Chechen Operation," in Glenn, ed., *Capital Preservation*, 2001, pp. 49–50.

[250]Andrei Viktorov, "Predpraznichniy Shturm [Preholiday Storm]," *Segodnya*, February 24, 2000; Vladimir Bochkarev and Vladimir Komol'tzev, "Rossiyskaya 'Burya v Gorakh' [Russian 'Mountain Storm']," *Nezavisimoye Voyennoye O bozreniye*, February 25, 2000; Boikov, "Luchshe Gor Mogut Bit'... [Better Than Mountains Could Be...]

Air-defense missiles were not often used against helicopters, but RPGs were used against helicopters that flew within range.[251]

At the top of the list of insights from the Russian experience is the vulnerability of many of their vehicles in the mountain environment. The Russians' lack of good information and the Chechens' excellent understanding of both the terrain and their enemy contributed both to the length of this conflict and to the high levels of Russian casualties.[252]

Difficult terrain typically provides an excellent illustration of the trade-offs between maneuverability and survivability. In Chechnya, however, the Russians had neither maneuverability nor survivability—their vehicles, armored and otherwise, were unable to reach many areas and also remained vulnerable to attack and destruction. The result was even heavier reliance on helicopters and fixed-wing aviation, but these too faced limits.

Stryker Brigade Combat Teams in Operation Iraqi Freedom (2003–2005)

DOTMLPF Insights. This case offers insights in the areas of doctrine, training, leader development, organization, and materiel.

Doctrine. The U.S. Army developed specific doctrine for the SBCTs, including dedicated manuals at every level (from the brigade to the individual soldier). SBCT doctrine is optimized for small-scale contingencies such as stability operations. It is built around the idea of defeating an adversary's decision cycle by "seeing first, understanding first, deciding first, and finishing decisively."[253] The modest level of

---

[251] Oliker, *Russia's Chechen Wars*, p. 71; Andrei Smolin and Viktor Kolomietz, "Khuzhe Gor Mogut Bit' Tol'ko Gori ... [Only Mountains Are Worse Than Mountains ...]," *Armeiskii Sbornik*, March 2000; Aleksander Bugai and Oleg Bedula, "Polyot Protiv Solntze [Flight Against the Sun]," *Krasnaya Zvezda*, May 10, 2000. Smolin and Kolomietz also report that the rebels had Stingers, but this is extremely unlikely.

[252] Casualty levels for Chechnya are notoriously difficult to estimate, but it seems clear that they were high.

[253] Roger M. Stevens and Kyle J. Marsh, "3/2 SBCT and the Countermortar Fight in Mosul," Field Artillery, No. PB6-05-1, January–February 2005, p. 37.

armor protection provided by the Stryker vehicles, compared to the M1 Abrams and M2 Bradley, make such an approach important.

Training. The SBCTs benefited immensely from the relatively low turnover rates of their personnel before being deployed to Iraq. Many 3/2 SBCT personnel joined the unit when it was established in 1999 and deployed with it to Iraq in 2003. Because the U.S. Army believed that the SBCT would need to be employed in new ways to be successful, it placed great emphasis on stabilizing and training the SBCTs. As a result, the SBCTs employed in OIF were cohesive and well trained.

Leader Development. The relatively low turnover rates of SBCT officers provided significant advantages in the area of leader development. Additionally, aggressive and innovative officers appear to have been drawn to the challenge of establishing a new type of brigade.

Organization. The unique SBCT organization appears to have been broadly successful. The relative dearth of armor protection on the Stryker increases the importance of integrated combined-arms tactics. The SBCT incorporates many capabilities, like field artillery and RSTA, that are traditionally pooled above the brigade level. Including these capabilities within the SBCT appears to have fostered tight combined-arms integration. The U.S. Army evidently agrees with this assessment, since it decided to "modularize" the U.S. Army structure, a move that essentially transfers the SBCT mode of organization to the rest of the U.S. Army's brigades.

Materiel. The Stryker was a successful combat vehicle in Iraq between 2003 and 2005. It provided better mobility, protection, armament, and sensors than an armored HMMWV. Most SBCTs are converted light-infantry brigades that would otherwise operate with armored HMMWVs; therefore, their conversion to the Stryker was clear improvement for these units in the context of the Iraqi threat environment.

The Stryker falls short of the M1 Abrams and M2 Bradley in terms of protection, armament, and sensors. However, these shortfalls were not operationally significant in the relatively limited threat environment in northern Iraq between 2003 and 2005.

There are a number of materiel issues with the Stryker vehicle. Without slat armor, the Stryker is clearly very vulnerable on the contemporary battlefield. While slat armor makes the vehicle much more difficult to destroy with high-explosive antitank warheads, it also makes it much heavier and more difficult to maneuver. The armor has also created problems with tire pressure, headlight positioning, and rear ramp functionality.

The RWS currently lacks stabilization for the weapon and the sensors. The imaging sensor has been criticized for its low magnification and field of view, and some argue that the thermal sensor is insufficiently sensitive. The flat-panel display currently operates in black and white, making target identification difficult. The RWS also lacks a laser designator, an obvious deficiency given the growing number and importance of laser-guided munitions.

Apart from the question of armor, however, these materiel issues are surprisingly minor given how recently the Stryker was added to the U.S. Army inventory. In fact, the Stryker performed well enough that the U.S. Army decided not to ship vehicles to and from Iraq between the first two SBCT rotations. The 1/25 simply assumed ownership of the 3/2 SBCT's vehicles in theater, indicating that the Stryker's shortcomings were sufficiently minor that there was no need for major adjustments before the 1/25 deployed. Moreover, the U.S. Army has moved quickly to address the various slat armor functionality issues previously described, and is developing a new, stabilized RWS that features better sensors and a laser designator.

Ultimately, though, the question of armor remains. A lack of full-spectrum armor protection is precisely what makes medium-weight vehicles what they are. The basic Stryker is vulnerable, and even slat armor leaves the wheel wells and top of the vehicle poorly protected. Fortunately, this did not compromise the operational effectiveness of the SBCT in Iraq between 2003 and 2005.

BOS Insights. This case yields BOS insights in the areas of maneuver, fire support, intelligence, mobility and survivability, and CSS.

Maneuver. SBCTs were used in a variety of roles, including presence patrolling, route security, cordon and search, and raids. In the Iraqi threat environment of this period, the Strykers provided an effective means of closing with and destroying the enemy.

Fire Support. The SBCTs employed relatively limited firepower in Iraq. Most of the field artillery was left stateside, and that which did deploy was employed largely in the countermortar and counterbattery mission. There is no evidence that the SBCT lacked fire support in Iraq between 2003 and 2005.

Intelligence. Accurate intelligence is critical to successful counterinsurgency and stability operations. The SBCT's enhanced technical and human intelligence capabilities appear to have been useful in this regard, as was the ability (provided by the All-Source Analysis System) to tap into national intelligence sources. The SBCTs were better-prepared than most other U.S. Army units for the intelligence requirements of operations in Iraq between 2003 and 2005.

Mobility and Survivability. The principal threat to SBCT survivability was the IED. As a relatively light vehicle, the Stryker is quite vulnerable to IEDs, particularly those that are vehicle-borne. SBCTs must particularly alert to this threat when they are deployed and the SBCT community must take this threat into account in future plans.

More importantly, the SBCT experience in Iraq between 2003 and 2005 sheds light on the degree to which improved situational awareness cannot substitute for traditional armor. The SBCTs possessed the most advanced digital battle command systems available to any army at the time, and through these systems the SBCT could access national-level sources of intelligence and information. Even with these capabilities, however, there is no evidence that the SBCT was able to significantly enhance its survivability through detailed situational awareness of the enemy's capabilities, intentions, and dispositions. The nature of the irregular adversary rendered much of this capability inapplicable. This suggests that while digital battle command systems are very useful, particularly in making planning and execution more rapid and precise, they cannot substitute for armor protection.

Combat Service Support. The Strykers are widely lauded for their supportability. Their ready rate from 2003 to 2005 exceeded 90 percent, and they required far less CSS than a comparable heavy unit. Because SBCTs are a modular brigade, however, the U.S. Army experienced some difficulty in arranging the CSS functions traditionally handled at the division level.[254]

Characteristics of Transformation Insights. This case yields transformation insights in the areas of agility, versatility, lethality, survivability, and sustainability. Note that sustainability insights are discussed above in the BOS section on CSS.

Agility. The SBCTs proved quite agile in Iraq between 2003 and 2005. Their road mobility and organic capabilities enabled theater commanders to use them as a rapid-reaction force. The primary limitation on Stryker agility is their lack of armor protection. The SBCT is inappropriate for any situation in which an adversary may possess capable antiarmor weapons.

Versatility. The broad suite of organic capabilities made the SBCT a tremendously versatile unit. The advanced planning capabilities provided by the ABCS and FBCB2 also enhanced the unit's versatility.

Lethality. The SBCTs were highly lethal in Iraq between 2003 and 2005. The units easily achieved lethal overmatch against the modestly armed adversaries present in their AO.

Survivability. Because Strykers are more survivable than armored HMMWVs, SBCTs are, broadly speaking, more survivable than light units. Because Strykers are less survivable than M1s and M2s, SBCTs are, broadly speaking, less survivable than heavy units.

In the context of Iraq between 2003 and 2005, the Strykers were sufficiently survivable. However, even a minor change in the threat environment, such as the adversary's acquisition of antitank guided missiles or medium-caliber antitank cannons, would have radically altered this picture.

Future commanders must take care to commit SBCTs to appropriate threat environments. They must also monitor the evolution of the threat to ensure that the SBCT does not become unduly vulnerable during the course of a deployment.

---

[254]"Transcript: Brig-Gen. Ham on the Stryker in Iraq," *Defense Industry Daily*, October 16, 2006, which reports an operational readiness of 94.66 percent. See also Robert Brown, Commander 1st Brigade, 25th Infantry Division, Multinational Force-Northwest, "Special Department of Defense Operational Update Briefing on Operations in Northwest Iraq," transcript, U.S. Department of Defense, September 14, 2005.

Complex Terrain Insights. The SBCTs operated in urban terrain in Iraq between 2003 and 2005. The Stryker vehicles provided a useful mix of capabilities for urban combat. Their level of protection was sufficient for operation in urban areas, and they were more mobile and agile in built-up areas than were heavy forces.

Medium-Armored Forces in Operations at the Lower End of the Range of Military Operations

The Rescue of Task Force Ranger in Mogadishu, Somalia (1993)

DOTMLPF Insights. This case offers insights in the areas of training and materiel.

Training. Task Force Ranger and the 10th Mountain Division QRF had never trained together or rehearsed extraction operations. Similarly, the Pakistani and Malaysian forces had never trained with the U.S. forces before the action on October 3–4. In fact, none of these forces were requested until after Task Force Ranger became trapped. Therefore, it took time to assemble the ad hoc rescue force and to prepare and brief the rescue plan to the participants. Finally, the Malaysian crews did not speak English, further complicating matters.

Materiel. The rescue of Task Force Ranger clearly demonstrates that medium-armored forces, because of their protected mobility and firepower, can make a critical difference during attempts to extract light and special operating forces from untenable positions. However, four Malaysian Condor APCs were lost to fire from Somali RPGs—weapons that are ubiquitous throughout the world.

BOS Insights. This case offers insights in the areas of maneuver, fire support, C2, and mobility and survivability.

Maneuver. The rescue of Task Force Ranger required maneuverability in the streets. Medium armor supplied this maneuverability in the form of Malaysian Condor APCs, which were supported by Pakistani M48 tanks. Two previous attempts by lightly armored and unarmored vehicles were repulsed by intense Somali fire.

Fire Support. Task Force Ranger, the 10th Mountain Division QRF, and the Pakistani and Malaysian forces used virtually every weapon at their disposal to extract the trapped elements of Task Force Ranger. In addition to small arms and grenades, these weapons included 7.62-mm miniguns, .50-caliber machine guns, 20-mm cannons, Mk-19 grenade launchers, 2.75-inch rockets, and TOW missiles. This massive firepower, coupled with the medium armor, enabled the extraction of the cut-off U.S. light forces.

Command and Control. C2 of the elements attempting to rescue Task Force Ranger proved difficult for many reasons. First, the rescue force found itself in an inherently chaotic environment that made C2 difficult. Second, the rescue force was composed of coalition forces that spoke different languages (English, Pashto, Punjabi, Urdu, Chinese, and Malay).[255] Third, the rescue force was ad hoc, and C2 relationships were established on the fly and in extremis. Fourth, U.S. soldiers in the relief convoy had night vision devices while the coalition forces did not. Finally, the soldiers of the 10th Mountain Division had difficulty using their radios from inside the Malaysian Condor APCs, "making communication nearly impossible and plaguing the soldiers throughout the mission."[256]

Mobility/Countermobility/Survivability. The barriers and roadblocks placed by the Somalis complicated maneuver for light vehicles and trucks. Pakistani tanks and Malaysian Condor APCs were able to maneuver much more effectively because they could break through these improvised barriers.

Characteristics of Transformation Insights. This case yields transformation insights in the areas of lethality and survivability.

Lethality. Part of the reason that the trapped elements of Task Force Ranger and its rescuers survived was the massive amount of firepower—provided by systems ranging from personal weapons to gunship-delivered ordnance—they could deliver to suppress and kill the Somali militia. Additionally, the Condor APC's firing ports enabled infantrymen within the vehicle to fire their personal weapons.

Survivability. Medium armor, particularly when combined with massive firepower, had sufficient survivability to maneuver during the

---

[255] Bolger, *Death Ground*, p. 221.

[256] Casper, *Falcon Brigade*, p. 59.

October 3–4 battle in Mogadishu. Light vehicles could not, because of their lack of armor and consequent vulnerability to even small arms fire, much less machine guns and RPGs.

Complex Terrain Insights. The October 3–4 battle in Mogadishu was an MOUT fight. Skilled infantrymen, protected by medium-armored vehicles, prevailed under a firestorm of firepower. Even so, the medium-armored force lost four vehicles to enemy RPGs. The battlefield advantage largely belonged to the Somalis, who could maneuver on familiar terrain and shoot down on U.S. and coalition vehicles from rooftops.

Australia and New Zealand in East Timor (1999–2000)

DOTMLPF Insights. This case offers insights in the areas of doctrine, training, organization, and materiel.

Doctrine. Before Operation Stabilise, the Australian Army focused on preparing for high-intensity operations. In General Cosgrove's words, "you learn to warfight and you adapt down for challenges for which outcomes are rendered more credible by your high-end skills."[257] Similarly, INTERFET validated the deployment of the armored vehicles normally reserved for high-intensity warfare in the minds of many:

> Before East Timor, there was a perception that the use of armour would likely escalate a conflict. However, the Australian Army's experiences . . . revealed that the introduction of armour prevented a confrontation from escalating and resolved it in Australia's favour, and generally without a need to fire.[258]

Lieutenant Colonel Krause, 2nd Cavalry, reported that, in East Timor, the Australians "saw something we don't see a lot of on exercise—we saw people scared of armour. We often talk about shock action, we

---

[257] Bostock, "East Timor: An Operational Evaluation," p. 27.

[258] Bostock, "East Timor: An Operational Evaluation," p. 27.

talk about the psychological effect of armour—we saw that in East Timor."[259]

Operation Stabilise also revealed deficiencies in Australian doctrine for conducting multinational operations. These deficiencies arose because the "possibility of Australia having to construct a multinational force had not been considered in advance of this operation; therefore no real doctrinal guidance existed."[260]

Training. The Australian and New Zealand soldiers and junior leaders were overwhelmingly regulars; only 3 percent of the Australians were reservists. These forces were well trained and disciplined and adhered to the rules of engagement, despite provocations.[261]

---

[259] Bostock, "East Timor: An Operational Evaluation," p. 27. See also, Sean M. Maloney, "Insights into Canadian Peacekeeping Doctrine," *Military Review*, Vol. 76, No. 2, March–April 1996, p. 20, which supports the "intimidation" value of armor in peacekeeping operations. The author notes the decisions made about the composition and experiences of the Canadian forces deployed during Operation Cavalier to support the expanded UN Protection Force (UNPROFOR) mission in Bosnia-Herzegovina in 1992:

> UN planners in New York wanted Canada to provide four light infantry companies with wheeled vehicles, one mechanized company with 15 armored personnel carriers (APCs) and 250 combat engineers. No heavy weapons such as a .50-caliber machineguns or mortars were to be taken along. Canadian planners generated a number of light force options but rapidly rejected all of them because of a high armor and artillery threat. The final organization was created based on the assumption that the contingent had to be prepared to defend itself against a ground assault which included tanks. . . . In all, Canada deployed a 900-man, four-company battalion group equipped with 83 M113 APCs, including eight (later 16) TOW Under Armor vehicles, four M113 mortar carriers and a 250-man armored engineer unit with combat engineer vehicles. . . . The TOWs possessed thermal imaging sights and were instrumental in providing the appropriate intimidation effects when the battalion group ran into belligerent roadblocks on the road to Sarajevo.

See also p. 21, where the value of medium forces in peacekeeping is further highlighted:

> Many local belligerent commanders, who were used to shooting up 'soft' UN humanitarian relief convoys, thought twice about interfering with convoys escorted by the Canadian battle group in Bosnia. By comparison [with Canadian medium forces], the other national contingents deploying with UNPROFOR were very light on the ground, with practically no wheeled transport or APCs.

[260] Ryan, *Primary Responsibilities and Primary Risks*, p. 119.

[261] Ryan, *Primary Responsibilities and Primary Risks*, pp. 71–72.

Organization. The Australians made an organizational modification for INTERFET. C Squadron, 2nd Cavalry Regiment, was organized for its mission of tactical reconnaissance. During Operation Stabilise, this unit's ASLAVs were used to "provide mobility, protection and communications" for the normally dismounted infantry of the 3rd Royal Australia Rifles.[262]

Materiel. Although they provided mobility and survivability to Australian and New Zealand forces, the armored vehicles deployed to East Timor did demonstrate shortcomings. The deficiencies of the M113 and ASLAV in the areas of lethality and survivability are discussed later in this section.

The M113s in the Australian and New Zealand armed forces were approximately 30 years old. Their deployment to East Timor highlighted the following deficiencies:

- an aging and maintenance-intensive power train and an obsolete steering and braking system that ran "hot" when the vehicle negotiated hills, bends, and corners
- a mix of old and new communications suites
- no effective shade protection for stationary vehicles
- no global positioning system (GPS) and no integrated tactical navigation system linked to the Battlefield Command Support System (BCSS)
- no effective wide field-of-view night-driving system
- no effective integrated AFV crewman ensemble.[263]

On a positive note, however, the M113 demonstrated outstanding off-road capabilities and an ability to "negotiate terrain that proved impassable to other vehicles, particularly in steep, confined terrain during the

[262]Bostock, "East Timor: An Operational Evaluation," p. 26.

[263]Bostock, "East Timor: An Operational Evaluation," p. 26.

monsoon season."[264] This cross-country mobility was due to the M113's "low ground pressure of 120kPa."[265]

The ASLAV's shortcomings included the absence of "a hybrid tactical navigation system with vehicle-integrated GPS" and the need for "a BCSS capability to improve commander and gunner situational awareness."[266] Furthermore, because it is a wheeled vehicle, the ASLAV has a relatively high ground pressure of 375 kPa, which caused some Australians to fear that the ASLAV would experience "extreme difficulty in carrying out reconnaissance off roads and tracks during the wet season."[267]

BOS Insights. Operation Stabilise offers insights in maneuver and CSS.

Maneuver. The protected mobility afforded by the M113s and LAVs made a significant contribution to the operational capabilities of INTERFET. The terrain in East Timor, however, highlighted the superior cross-country capability of tracked vehicles, with their low ground pressure. ASLAV were generally confined to roads because of their higher ground pressure.[268]

Combat Service Support. Australia experienced some difficulty in supporting its deployed armored force. Units experienced shortages of spare parts and consumables once the supplies they brought with

---

264 Bostock, "East Timor: An Operational Evaluation," p. 26.

265 John R. Lenehan, "The Impact of the White Paper on Australian Armour," *Defender*, Vol. VXIII, No. II, Winter 2001, p. 8.

266 Bostock, "East Timor: An Operational Evaluation," p. 27.

267 Lenehan, "The Impact of the White Paper on Australian Armour," p. 8. See also Australian Department of Defense, "LAND 106-M113 Upgrade Project," November 1, 2007; Australian Department of Defense, Defence Materiel Organization, *Current and Future Simulation Projects, 2002–2010*, May 2002; Australian Department of Defence, "Equipment: Light Armored Vehicle," n.d. Australia is upgrading approximately 350 of its M113s, giving them greater reliability, mobility, survivability, and lethality. New Zealand, on the other hand, is replacing its M113 fleet with a Canadian-built LAV-III variant called the NZLAV.

268 See Paul Hornback, "The Wheel Versus Track Dilemma," *Armor*, Vol. 107, No. 2, March–April 1998, p. 33. Hornback notes the that "[g]enerally as vehicle ground pressure increases, cross country trafficability decreases," and that "when the gross vehicle weight exceeds 20 tons and off-road usage remains above 60 percent, a tracked configuration is required to guarantee the best mobility for unrestricted, all-weather tactical operations."

them were depleted. Ironically, the "push method" of sending units supplies according to preoperational plans occasionally delivered too much of the wrong thing to units.[269] The fact that Australia found itself supporting the majority of INTERFET was a large contributing factor. Australia's "policy of holding minimum war stocks of supplies and provisions and operation on the 'just in time' principle" also contributed to logistics difficulties.[270]

Characteristics of Transformation Insights. This case yields transformation insights in the areas of responsiveness, deployability, agility, versatility, lethality, survivability, and sustainability.

Responsiveness. The Australian and New Zealand components of INTERFET were able to respond rapidly to the crisis in East Timor with highly capable forces that were supported by medium-armored vehicles. Australia made the conscious decision, based on the low threat to medium-armored vehicles and the requirement to respond rapidly, to not deploy its Leopard MBTs. In a higher-intensity crisis, the Leopards would probably have been deployed; the Australian Army believes that the deployment of these MBTs would have slowed their response.

Deployability. The medium armor in the Australian and New Zealand armies made the initial contingents of INTERFET highly deployable. M113s deployed by C-130 transports and the *Jervis Bay* and *Tobruk* quickly brought other forces and vehicles.

Agility. The ADF were normally focused on preparing for high-intensity operations. Although soldiers were well trained and disciplined and led by very professional junior leaders, there were shortfalls. These included a lack of civil affairs capabilities and insufficient doctrine and preparation to lead a diverse, large-scale multinational coalition.[271] On

[269]Lieutenant Michael Krause (Australian Army, 2nd Cavalry Regiment), discussion with Fred Bowden, Australian Liaison at RAND Corporation, May 9, 2002.

[270]Bostock, "East Timor: An Operational Evaluation," p. 27.

[271]Ryan, *Primary Responsibilities and Primary Risks*, pp. 67, 110; see also p. 18, where General Cosgrove, the INTERFET commander, later argued against any moves to "lighten up the force structure to specialise in the sort of 'policing plus' role which typifies most peacekeeping missions." He believed that

the whole, however, their agility enabled them to make the transition from warfighting to peace enforcement with little difficulty.

Versatility. As previously mentioned, the cavalry troop, which had trained to perform tactical reconnaissance, used its ASLAVs to support infantry operations during Operation Stabilise. This showed the versatility of the unit and the equipment.

Lethality. Operations in East Timor revealed significant shortcomings in the M113 weapons station:

> The weapon station performed poorly. The vibrations when the machine guns fired did not allow the sight to be used, and the hatch could not be closed due the toxic fumes generated when the guns were fired. The vehicle today has neither a day nor a night sight. The weapons are not stabilised and accurate fire whilst moving is not possible, and the capability of neutralizing or destroying the enemy is low.[272]

Survivability. Although there were few INTERFET casualties during operations in East Timor, the deployment did reveal survivability problems in both the M113 and the ASLAV. Both proved vulnerable to small arms with armor-piercing ammunition, heavy machine guns, and RPGs. Furthermore, on the personnel-carrier version of the ASLAV, the .50-caliber machine gun's lack of a shield exposed the gunner to fire.[273] Australian soldiers commented on the vulnerability of the ASLAVs:

> ASLAV crews felt vulnerable to enemy small-arms fire and then-unknown anti-armour threats during the early stages; some said

---

> forces structured and equipped, ready if necessary, for war were actually very effective, probably more effective than had they been less capable. . . . A force optimised for peacekeeping would have in my view invited more adventurist behaviour by our adversaries.

[272]Lenehan, "The Impact of the White Paper on Australian Armour," p. 8.

[273]Lenehan, "The Impact of the White Paper on Australian Armour," pp. 8–9; Parliament of Australia, Joint Standing Committee on Foreign Affairs, Defence and Trade, "Technology, Equipment and Supplies," in *From Phantom to Force: Towards a More Efficient and Effective Army*, September 4, 2000, Chapter Eight.

> they would have preferred to have been operating with a troop of Leopard AS1 main battle tanks on landing at Dili.[274]

Sustainability. The CSS difficulties encountered by Australian forces in East Timor have previously been discussed. It is also important to note the contribution of sealift to the sustainment of INTERFET: "Over 90% of military cargo and people went into and out of East Timor by sea and the lack of roads and infrastructure meant that sea transport was vital in-theatre as well."[275]

Complex Terrain Insights. Although there are urban areas in East Timor, little fighting occurred in those areas. The principal complex terrain types that presented problems to INTERFET were the jungle and difficult off-road areas. As previously noted, M113s, with their low ground pressure, fared best in cross-country operations that were inaccessible to wheeled vehicles.

---

[274] Bostock, "East Timor: An Operational Evaluation," p. 26.

[275] Ryan, *Primary Responsibilities and Primary Risks*, p. 78.

APPENDIX C

Definitions

This appendix provides definitions used in this monograph for DOTMLPF, BOS, and characteristics of a transformed force.

DOTMLPF

The following DOTMLPF definitions are taken from the Chairman of the Joint Chiefs of Staff, Instruction 3170.01E, "Joint Capabilities Integration and Development System."[1]

Doctrine

> Fundamental principles that guide the employment of US military forces in coordinated action toward a common objective. Though neither policy nor strategy, joint doctrine serves to make US policy and strategy effective in the application of US military power. Joint doctrine is based on extant capabilities. Joint doctrine is authoritative guidance and will be followed except when, in the judgment of the commander, exceptional circumstances dictate otherwise.

[1] Chairman of the Joint Chiefs of Staff, Instruction 3170.01E, "Joint Capabilities Integration and Development System," pp. GL-9–GL-10.

Organization

A [joint] unit or element with varied functions enabled by a structure through which individuals cooperate systematically to accomplish a common mission and directly provide or support [joint] warfighting capabilities. Subordinate units/elements coordinate with other units/elements and, as a whole, enable the higher-level [joint] unit/element to accomplish its mission. This includes the joint manpower (military, civilian and contractor support) required to operate, sustain and reconstitute joint warfighting capabilities.

Training

Military training based on joint doctrine or joint tactics, techniques and procedures to prepare joint forces and/or joint staffs to respond to strategic and operational requirements deemed necessary by combatant commanders to execute their assigned missions. Joint training involves forces of two or more Military Departments interacting with a combatant commander or subordinate joint force commander; involves joint forces and/or joint staffs; and is conducted using joint doctrine or joint tactics, techniques and procedures.

Materiel

All items (including ships, tanks, self-propelled weapons, aircraft, etc., and related spares, repair parts and support equipment, but excluding real property, installations and utilities) necessary to equip, operate, maintain and support [joint] military activities without distinction as to its application for administrative or combat purposes.

Leadership and Education

Professional development of the joint commander is the product of a learning continuum that comprises training, experience, education and self-improvement. The role of Professional Military Education and Joint Professional Military Education is to provide the education needed to complement training, experience and self-improvement to produce the most professionally competent individual possible.

Personnel

The personnel component primarily ensures that qualified personnel exist to support joint capabilities. This is accomplished through synchronized efforts of joint force commanders and Service components to optimize personnel support to the joint force to ensure success of ongoing peacetime, contingency and wartime operations.

Facilities

Real property consisting of one or more of the following: a building, a structure, a utility system, pavement and underlying land. Key facilities are selected command installations and industrial facilities of primary importance to the support of military operations or military production programs. A key facilities list is prepared under the policy direction of the Joint Chiefs of Staff.

Battlefield Operating Systems

The following definitions of a BOS and the BOS types are taken from U.S. Department of the Army, FM 3-0, *Operations*.[2]

---

[2] U.S. Department of the Army, FM 3-0, *Operations*, pp. 5-15–5-18.

The Battlefield Operating System Concept

Armed with a coherent and focused intent, commanders and staffs develop the concept of operations and synchronize the BOS. The BOS are the physical means (soldiers, organizations, and equipment) used to accomplish the mission. The BOS group related systems together according to battlefield use.

Intelligence

The intelligence system plans, directs, collects, processes, produces, and disseminates intelligence on the threat and environment to perform intelligence preparation of the battlefield (IPB) and the other intelligence tasks. A critical part of IPB involves collaborative, cross-BOS analysis across echelons and between analytic elements of a command. The other intelligence tasks are—

- Situation development.
- Target development and support to targeting.
- Indications and warning.
- Intelligence support to battle damage assessment.
- Intelligence support to force protection.

Intelligence is developed as a part of a continuous process and is fundamental to all Army operations.

Maneuver

Maneuver systems move to gain positions of advantage against enemy forces. Infantry, armor, cavalry, and aviation forces are organized, trained, and equipped primarily for maneuver. Commanders maneuver these forces to create conditions for tactical and operational success. By maneuver, friendly forces gain the ability to destroy enemy forces or hinder enemy movement by direct and indirect application of firepower, or threat of its application.

Fire Support

Fire support consists of fires that directly support land, maritime, amphibious, and special operations forces in engaging enemy forces, combat formations, and facilities in pursuit of tactical and operational objectives. Fire support integrates and synchronizes fires and effects to delay, disrupt, or destroy enemy forces, systems, and facilities. The fire support system includes the collective and coordinated use of target acquisition data, indirect-fire weapons, fixed-wing aircraft, electronic warfare, and other lethal and nonlethal means to attack targets. At the operational level, maneuver and fires may be complementary in design, but distinct in objective and means.

Air Defense

The air defense system protects the force from air and missile attack and aerial surveillance. It prevents enemies from interdicting friendly forces while freeing commanders to synchronize maneuver and firepower. All members of the combined arms team perform air defense tasks; however, ground-based air defense artillery units execute most Army air defense operations. These units protect deployed forces and critical assets from observation and attack by enemy aircraft, missiles, and unmanned aerial vehicles. The WMD [weapons of mass destruction] threat and proliferation of missile technology increase the importance of the air defense system. Theater missile defense is crucial at the operational level.

Mobility/Countermobility/Survivability

Mobility operations preserve friendly force freedom of maneuver. Mobility missions include breaching obstacles, increasing battlefield circulation, improving or building roads, providing bridge and raft support, and identifying routes around contaminated areas. *Countermobility* denies mobility to enemy forces. It limits the maneuver of enemy forces and enhances the effectiveness of fires. Countermobility missions include obstacle building and smoke generation. *Survivability* operations protect friendly forces from the effects of enemy weapons systems and from natural occurrences. Hardening of facilities and fortification of battle positions are active survivability measures. Military deception, OPSEC, and dispersion can also increase survivability. NBC [nuclear, biological, and chemical] defense measures are essential survivability tasks.

Combat Service Support

CSS includes many technical specialties and functional activities. It includes the use of host nation infrastructure and contracted support. CSS provides the physical means for forces to operate, from the production base and replacement centers in the continental US to soldiers engaged in close combat. It is present across the range of military operations, at all levels of war.

Command and Control

Command and control has two components—the commander and the C2 system. Communications systems, intelligence systems, and computer networks form the backbone of C2 systems and allow commanders to lead from any point on the battlefield. The C2 system supports the commander's ability to make informed decisions, delegate authority, and synchronize the BOS. Moreover, the C2 system supports the ability of commanders to adjust plans for future operations, even while focusing on the current fight. Staffs work within the commander's intent to direct units and control resource allocations. They also are alert to spotting enemy or friendly situations that require command decisions and advise commanders concerning them. Through C2, commanders initiate and integrate all military functions and systems toward a common goal: mission accomplishment . . . .

Reliable communications are central to C2 systems. Effective battle command requires reliable signal support systems that

> enable commanders to conduct operations at varying tempos. Nonetheless, commanders, not their communication systems, dictate command style. Signal planning increases the commander's options by providing signal support to pass vital information at critical times. This capability allows commanders to leverage tactical success and anticipate future operations. Communications planning is a vital component of maintaining or extending operational reach.

Characteristics of a Transformed Force

The following required attributes of a transformed force are taken from General Shinseki's October 21, 1999, statement to the House Armed Services Committee on the status of forces.[3]

Responsive

> Responsiveness has the quality of time, distance, and sustained momentum. Our threat of the use of force, if it deters miscalculation by adversaries, provides a quality of responsiveness all its own. We will provide strategic responsiveness through forward-deployed forces, forward positioned capabilities, engagement, and, when called, through force projection from the CONUS [continental United States] or any other location where needed capabilities reside. Wherever soldiers serve, we are part of the Nation's solution to its tremendous world leadership responsibilities.

[3] U.S. House Armed Services Committee, "Statement by General Eric K. Shinseki, Chief of Staff, United States Army, on Status of Forces," 106th Congress, October 12, 1999. See also U.S. Department of the Army, *Army Transformation Wargame 2001*, pp. 2–3; U.S. Department of the Army, FM-1, *The Army*, 2005, p. 4-3, which lists the characteristics of Army transformation, but does not define them.

Deployable

We will develop the capability to put combat force anywhere in the world in 96 hours after liftoff—in brigade combat teams for both stability and support operations and for warfighting. We will build that capability into a momentum that generates a warfighting division on the ground in 120 hours and five divisions in 30 days.

Agile

We will attain the mental and physical agility operationally to move forces from stability and support operations to warfighting and back again just as we have demonstrated the tactical warfighting agility to task organize on the move and transition from the defense to the offense and back again. We will develop leaders at all levels and in all components who can prosecute war decisively and who can negotiate and leverage effectively in those missions requiring engagement skills.

Versatile

We will design into our organizational structures, forces which will, with minimal adjustment and in minimum time, generate formations which can dominate at any point on the spectrum of operations. We will also equip and train those organizations for effectiveness in any of the missions that The Army has been asked to perform. These commitments will keep our components capable, affordable, and indispensable to the Nation.

Lethal

The elements of lethal combat power remain fires, maneuver, leadership, and protection. When we deploy, every element in the warfighting formation will be capable of generating combat power and contributing decisively to the fight. We will retain today's light force deployability while providing it the lethality and mobility for decisive outcomes that our heavy forces currently enjoy. We will retain heavy force lethality through overmatch while giving it deployability and employability in areas currently accessible only by light forces. We intend to get to trouble spots faster than our adversaries can complicate the crisis, encourage de-escalation through our formidable presence, and if deterrence fails, prosecute war with an intensity that wins at least cost to us and our allies and sends clear messages to all who threaten America. As technology allows, we will begin to erase the distinctions between heavy and light forces. We will review our requirement for specialty units and ensure they continue to evolve to meet the needs of the Nation.

Survivable

We will derive the technology that provides maximum protection to our forces at the individual soldier level whether that soldier is dismounted or mounted. Ground and air platforms will leverage the best combination of low observable, ballistic protection, long range acquisition and targeting, early attack, and higher first round hit and kill technologies at smaller calibers that are available. We are prepared to venture into harm's way to dominate the expanded battlespace, and we will do what is necessary to protect the force.

Sustainable

We will aggressively reduce our logistics footprint and replenishment demand. This will require us to control the numbers of vehicles we deploy, leverage reach back capabilities, invest in a systems approach to the weapons and equipment we design, and revolutionize the manner in which we transport and sustain our people and materiel. We are prepared to move to an all wheel formation as soon as technology permits.

Bibliography

"3-2 SBCT Arrowhead Brigade Capabilities Overview," briefing, February 2006.

Aldis, Anne, ed., *Strategic and Combat Studies Institute Occasional Paper No. 40: The Second Chechen War*, Shrivenham, UK: Conflict Studies Research Centre, 2000.

Alexander, McGill, "An African Rapid-Deployment Force." *Military Review*, Vol. 77, No. 3, May–June 1997, pp. 26–32.

Alexiev, Alex, *Inside the Soviet Army in Afghanistan*, Santa Monica, Calif.: RAND Corporation, R-3627-A, 1988. As of February 5, 2008:
http://www.rand.org/pubs/reports/R3627/

———, and Robert Nurick, *The Soviet Military Under Gorbachev: Report on a RAND Workshop*, Santa Monica, Calif.: RAND Corporation, R-3907-RC, 1990. As of February 5, 2008:
http://www.rand.org/pubs/reports/R3907/

Alexievich, Svetlana, *Zinky Boys: Soviet Voices from the Afghanistan War*, trans. Julia Whitby and Robin Whitby, New York: W. W. Norton, 1992.

Allard, Kenneth, *Somalia Operations: Lessons Learned*, Fort McNair, Washington, D.C.: National Defense University Press, 1995.

———, "Soviet Airborne Forces and Preemptive Power Projection," *Parameters*, Vol. 10, No. 4, December 1980, pp. 42–51.

Antonov, Sergey, "Armiya Gotovitsa Vikurivat' Boyevikov [Army Prepares to Smoke Out Rebels]," *Segodnya*, February 14, 2000.

"Armoured Personnel Carries (Wheeled), South Africa," *Jane's Armour and Artillery 2001–2002*, n.d.

Aseev, A., and A. Malyugin, "Na Tryokh Kitakh [On Three Whales]," *Armeiskii Sbornik*, November 2000.

Association of the U.S. Army, *Army Green Book 2005*, Washington, D.C., 2005.

Atkinson, Rick, "The Raid That Went Awry," *Washington Post*, January 30, 1994, p. A1.

———, "Night of a Thousand Casualties," *Washington Post*, January 31, 1994, p. A1.

Australian Department of Defence, "Equipment: Light Armored Vehicle," n.d. As of July 17, 2001:
http://www.defence.gov.au/army/equipment/aslav.htm

———, "Current Equipment–ASLAV Family," 2001. As of July 19, 2001:
http://www.defence.gov.au/army/2cav/newgear.html

———, Defence Materiel Organization, *Current and Future Simulation Projects, 2002–2010*, May 2002. As if February 5, 2008:
http://www.defence.gov.au/dmo/id/publications/CFSP_02.pdf

———, "LAND 112–ASLAV (Australian Light Armoured Vehicle)," September 21, 2007. As of February 5, 2008:
http://www.defence.gov.au/dmo/lsd/land112/land112.cfm

———, "LAND 106-M113 Upgrade Project," November 1, 2007. As of February 5, 2008:
http://www.defence.gov.au/dmo/lsd/land106/land106.cfm

Australian Minister for Defence, "Timor UN Resolution CNBC," September 15, 1999. As of February 5, 2008:
http://www.minister.defence.gov.au/1999/mt0899.htm

Baily, Charles M., *Faint Praise: American Tanks and Tank Destroyers During World War II*, Hamden, Conn.: Archon Books, 1983.

———, "Tank Myths," *Armor*, Vol. 110, No. 5, September–October 2001, pp. 36–38.

Barker, Anne, "Australia/Indonesia," Voice of America, correspondent report no. 2-299043, Canberra, Australia, May 11, 2000. As of February 5, 2008:
http://www.fas.org/irp/news/2000/05/000511-intell.htm

Beevor, Antony, *The Battle for Spain: The Spanish Civil War, 1936–1939*, New York: Penguin Books, 2006.

Bell, Raymond E., Jr., "Evolving Army Armor Structure in the Late 1920s," *Armor*, Vol. 110, No. 4, July–August 2001, pp. 29–34, 37.

Belosludtzev, Oleg, "Variation in Tactics of Actions," *Nezavisimoye Voyennoye Obozreniye*, May 12, 2000.

Bland, Larry I., Joellen K. Bland, and Sharon Ritenour Stevens, eds., *George C. Marshall Interviews and Reminiscences for Forrest C. Pogue*, 3rd ed., Lexington, Va.: George C. Marshall Research Foundation, 1996.

Blaxland, John, Major (Australian Army), email discussion with Charlie Cromwell on the experience of Australian forces in the East Timor peacekeeping operation, July 29, 2001.

Blumenson, Martin, *Breakout and Pursuit*, Washington, D.C.: Center of Military History, U.S. Army, [1961] 1984.

Bochkarev, Vladimir, and Vladimir Komol'tzev, "Rossiyskaya 'Burya v Gorakh' [Russian 'Mountain Storm']," *Nezavisimoye Voyennoye O bozreniye*, February 25, 2000.

Boikov, Roman, "Luchshe Gor Mogut Bit' . . . [Better Than Mountains Could Be . . .]," *Krasnaya Zvezda*, March 17, 2000.

Bolger, Daniel P., *Savage Peace: Americans at War in the 1990s*, Novato, Calif.: Presidio Press, 1995.

———, *Death Ground: Today's American Infantry in Battle*, Novato, Calif.: Presidio Press, 2000.

Boot, Max, *The Savage Wars of Peace: Small Wars and the Rise of American Power*, New York: Basic Books, 2002.

Bostock, Ian, "East Timor: An Operational Evaluation," *Jane's Defense Weekly*, Vol. 33, No. 18, May 3, 2000, pp. 23–27.

Bowden, Mark, *Black Hawk Down: A Story of Modern War*, New York: Atlantic Monthly Press, 1999.

Breen, Bob, "INTERFET: Some Wake Up Calls for the Australian Defence Force," *Defender*, Vol. XVIII, No. 2, Winter 2001, pp. 25–29.

Bridgland, Fred, *The War for Africa: Twelve Months that Transformed a Continent*, Gibraltar: Ashanti Publishers, 1991.

Brodie, Bernard, and Fawn Brodie, *From Crossbow to H-Bomb: The Evolution of the Weapons and Tactics of War*, rev. ed., Bloomington, Ind.: Indiana University Press, 1975.

Brown, Robert, Commander 1st Brigade, 25th Infantry Division, Multinational Force-Northwest, "Special Department of Defense Operational Update Briefing on Operations in Northwest Iraq," transcript, U.S. Department of Defense, September 14, 2005. As of February 8, 2008: http://www.defenselink.mil/Transcripts/Transcript.aspx?TranscriptID=2109

Bugai, Aleksander, and Oleg Bedula, "Polyot Protiv Solntze [Flight Against the Sun]," *Krasnaya Zvezda*, May 10, 2000.

CALL—*see* Center for Army Lessons Learned.

Cameron, Robert S., "Americanizing the Tank: U.S. Army Administration and Mechanized Development within the Army, 1917–1943," dissertation, Philadelphia: Temple University, 1994.

———, "Armor Combat Development 1917–1945," *Armor*, Vol. 106, No. 5, September–October 1997, pp. 14–19.

Candil, Antonio J., "Soviet Armor in Spain: Aid Mission to Republicans Tested Doctrine and Equipment," *Armor*, Vol. 108, No. 2, March–April 1999, pp. 31–38.

Cash, John A., John Albright, and Allan W. Sandstrum, *Seven Firefights in Vietnam*, Washington, D.C.: U.S. Office of the Chief of Military History, U.S. Army, [1970] 1989.

Casey, Ken, "Urban Combat in World War II: How Doctrine Changed as the War Progressed," *Armor*, Vol. 108, No. 6, November–December 1999, pp. 8–13.

Casper, Lawrence E., *Falcon Brigade: Combat and Command in Somalia and Haiti*, Boulder, Colo.: Lynne Rienner Publishers, 2001.

Celestan, Gregory J., "Wounded Bear: The Ongoing Russian Military Operation in Chechnya," Fort Leavenworth, Kan.: Foreign Military Studies Office, August 1996.

Center for Army Lessons Learned, *Operation Just Cause Lessons Learned:* Vol. I, *Soldiers and Leadership* (No. 90-9), Fort Leavenworth, Kan.: U.S. Army Combined Arms Command, 1990.

———, *Operation Just Cause Lessons Learned:* Vol. II, *Operations* (No. 90-9), Fort Leavenworth, Kan.: U.S. Army Combined Arms Command, 1990.

———, *Operation Just Cause Lessons Learned:* Vol. III, *Intelligence, Logistics & Equipment* (No. 90-9), Fort Leavenworth, Kan.: U.S. Army Combined Arms Command, 1990.

Center for Strategic Studies: New Zealand, "Strategic and Military Lessons from East Timor," *Center for Strategic Studies Briefing Papers*, Vol. 2, Part 1, February 2000. As of February 5, 2008:
http://www.victoria.ac.nz/css/docs/Strategic_Briefing_Papers/Vol.2%20Feb%20
2000/East%20Timor.pdf

Central Intelligence Agency, "OSS in Asia," Center for the Study of Intelligence Publications, November 19, 2007. As of February 5, 2008:
https://www.cia.gov/library/center-for-the-study-of-intelligence/csi-publications/
books-and-monographs/oss/art09.htm

Chalk, Peter, *Australian Foreign and Defense Policy in the Wake of the 1999/2000 East Timor Intervention*, Santa Monica Calif.: RAND Corporation, MR-1409-SRF, 2001. As of February 5, 2008:
http://www.rand.org/pubs/monograph_reports/MR1409/

"Challenges and Opportunities for Increments II and III Future Combat Systems (FCS)," Army Science Board, Summer 2003.

Chairman of the Joint Chiefs of Staff, Instruction 3170.01E, "Joint Capabilities Integration and Development System," May 11, 2005.

———, Instruction 3170.01F, "Joint Capabilities Integration and Development System," May 1, 2007.

Chamberlain, Peter, and Hilary Doyle, *Encyclopedia of German Tanks of World War Two: A Complete Illustrated Directory of German Battle Tanks, Armoured Cars, Self-Propelled Guns and Semi-Tracked Vehicles, 1933–1945*, London: Arms and Armour Press, [1978] 2001.

Chandler, Alfred D., Jr., Stephen E. Ambrose, Joseph P. Hobbs, Edwin Alan Thompson, and Elizabeth F. Smith, eds., *The Papers of Dwight David Eisenhower*, Baltimore: Johns Hopkins Press, [1970] 1971.

Charlston, Jeff, "The Evolution of the Stryker Brigade–From Doctrine to Battlefield Operations in Iraq," in McGrath, ed., 2006.

Cilliers, Jakkie, and Bill Sass, eds., *Mailed Fist: Developments in Modern Armour*, Institute for Security Studies, Monograph No. 2, March 1996.

Clarke, Jeffrey J., and Robert Ross Smith, *Riviera to the Rhine*, Washington, D.C.: Center of Military History, U.S. Army, 1993.

Cole, Hugh M., *The Lorraine Campaign*, Washington, D.C.: Center of Military History, U.S. Army, [1950] 1981.

———, *The Ardennes: Battle of the Bulge*, Washington, D.C.: Center of Military History, U.S. Army, [1965] 1988.

Cole, Ronald H., *Operation Just Cause: The Planning and Execution of Joint Operations in Panama, February 1988–January 1990*, Washington, D.C.: Joint History Office, 1995.

Collier, Craig A., "A New Way to Wage Peace: US Support to Operation Stabilise," *Military Review*, Vol. 81, No. 1, January–February 2001, pp. 2–9.

Commander, Army Ground Forces, "Memorandum for CSA," January 23, 1943, cited in Steadman, 1982.

Cordesman, Anthony H., and Abraham R. Wagner, *The Lessons of Modern War:* Vol. III, *The Afghan and Falklands Conflicts*, Boulder, Colo.: Westview Press, 1990.

———, *The Lessons of Modern War:* Vol. IV, *The Gulf War*, Boulder, Colo.: Westview Press, 1996.

Cooling, Benjamin Franklin, ed., *Case Studies in the Development of Close Air Support*, Washington, D.C.: Office of Air Force History, 1990.

Cooling, Norman L., "LAI in the MEU(SOC)," *Marine Corps Gazette*, Vol. 75, No. 8, August 1991, pp. 20–24.

Crile, George, *Charlie Wilson's War: The Extraordinary Story of the Largest Covert Operation in History*, New York: Atlantic Monthly Press, 2003.

Crocker, Chester A., "Southern African Peace-Making," *Survival*, Vol. 32, No. 3, May/June 1990, pp. 221–232.

———, *High Noon in Southern Africa: Making Peace in a Rough Neighborhood*, New York: W.W. Norton, 1992.

Crow, Duncan, and Robert J. Icks, *Encyclopedia of Tanks*, Secaucus, N.J.: Chartwell, 1975.

Cureton, Charles H., *United States Marine Corps in the Persian Gulf, 1990–1991: With the 1st Marine Division in Desert Shield and Desert Storm*, Washington, D.C.: Headquarters, U.S. Marine Corps, 1993.

Daley, John L. S., "The Theory and Practice of Armored Warfare in Spain: October 1936–February 1937," *Armor*, Vol. 108, No. 2, March–April 1999, pp. 30, 39–43.

———, "Soviet and German Advisors Put Doctrine to the Test: Tanks in the Siege of Madrid," *Armor*, Vol. 108, No. 3, May–June 1999, pp. 33–37.

Dastrup, Boyd L., *King of Battle: A Branch History of the U.S. Army's Field Artillery*, Fort Monroe, Va.: Office of the Command Historian, U.S. Army Training and Doctrine Command, 1993.

Day, Clifford E., "Critical Analysis on the Defeat of Task Force Ranger," thesis, Maxwell-Gunter Air Force Base, Ala.: Air Command and Staff College, 1997.

Dippenaar, J. M., "Armour in the African Environment," in Cilliers and Sass, eds., 1996. As of February 5, 2008:
http://www.iss.co.za/Pubs/Monographs/No2/Dippenaar.html

Donnelly, Thomas, Margaret Roth, and Caleb Baker, *Operation Just Cause: The Storming of Panama*, New York: Lexington Books, 1991.

Doubler, Michael D., *Closing With the Enemy: How GIs Fought the War in Europe, 1944–1945*, Lawrence, Kan.: University of Kansas Press, 1994.

Dunnigan, James F., and Raymond M. Macedonia, *Getting It Right: American Military Reforms After Vietnam to the Gulf War and Beyond*, New York: W. Morrow and Co., 1993.

Ellis, Chris, and Peter Chamberlain, *The Great Tanks*, London: Hamlyn Publishing Group, 1975.

"FAPLA—Angola's Marxist Armed Forces," *Jane's Soviet Intelligence Review*, July 1990, pp. 306–310.

Federation of American Scientists, "Military Analysis Network—U.S. Land Warfare Systems," n.d. As of February 5, 2008:
http://www.fas.org/man/dod-101/sys/land/

———, "Military Analysis Network—ZSU-23-4 Shilka 23MM Antiaircraft Gun," January 22, 1999. As of February 5, 2008:
http://www.fas.org/man/dod-101/sys/land/row/2s6m.htm

———, "Military Analysis Network—2S6M Tunguska Anti-Aircraft Artillery," June 19, 1999. As of February 5, 2008:
http://www.fas.org/man/dod-101/sys/land/row/2s6m.htm

Finch, Raymond C., III, "A Face of Future Battle: Chechen Fighter Shamil Basayev," *Military Review*, Vol. 77, No. 3, May–June 1997, pp. 33–41. As of February 5, 2008:
http://leav-www.army.mil/fmso/documents/shamil/shamil.htm

———, "Why the Russian Military Failed in Chechnya," Foreign Military Studies Office, Special Study No. 98-16, 1998. As of February 5, 2008:
http://leav-www.army.mil/fmso/documents/yrusfail/yrusfail.htm

Fleischer, Wolfgang, *Russian Tanks and Armored Vehicles, 1917–1945*, Atglen, Pa.: Schiffer, 1999.

———, and Richard Eiermann, *German Anti-Tank (Panzerjäger) Troops in World War II*, Atglen, Pa.: Schiffer Military History, 2004.

Fontenot, Gregory, E. J. Degen, and David Tohn, *On Point: The United States Army in Operation Iraqi Freedom*, Annapolis, Md.: Naval Institute Press, 2005.

Forty, George, *U.S. Army Handbook, 1939–1945*, Gloucestershire, UK: Sutton, 1997.

Foss, Christopher F., *Jane's World Armoured Fighting Vehicles*, New York: St. Martin's Press, 1976.

———, *Jane's Tanks and Combat Vehicles Recognition Guide*, New York: Harper Collins, 2000.

Fourth New Zealand Battery Group, "Country Profile—East Timor," n.d. As of July 17, 2001:
http://www.2lfg.mil.nz/timprof.htm

Friedman, Lawrence, *Kennedy's Wars: Berlin, Cuba, Laos, and Vietnam*, New York: Oxford University Press, 2000.

Frost, Matthew, "Czech Republic: A Chronology of Events Leading to the 1968 Invasion," RadioFreeEurope/RadioLiberty, August 20, 1998. As of February 5, 2008:
http://www.rferl.org/features/1998/08/F.RU.980820113706.asp

Gabel, Christopher R., *Seek, Strike, and Destroy: U.S. Army Tank Destroyer Doctrine in World War II*, Fort Leavenworth, Kan.: Combat Studies Institute, U.S. Army Command and General Staff College, 1985.

———, *The U.S. Army GHQ Maneuvers of 1941*, Washington, D.C.: Center of Military History, U.S. Army, 1991.

———, "World War II Armor Operations in Europe," in Hofmann and Starry, eds., 1999.

Gall, Carlotta, and Thomas de Waal, *Chechnya*, New York: New York University Press, 1998.

Geibel, Adam, "The Final Score: Russian Armor Losses in Chechnya Reflect Lethality of an Urban Fight," *Armor*, Vol. 108, No. 3, May–June 1999, p. 47.

———, "Ambush at Serzhen Yurt: Command-Detonated Mines in the Second Chechen War," *Engineer*, Vol. 31, No. PB 5-01-1, February 2001. As of February 5, 2008:
http://findarticles.com/p/articles/mi_m0FDF/is_1_31/ai_78974282

———, "Some Russian Tankers' Experiences In the Second Chechen War," *Armor*, Vol. 110, No. 4, July–August 2001, pp. 25–28.

The General Board, United States Forces, European Theater, *Organization, Equipment, and Tactical Employment of Separate Tank Battalions,* Study Number 50, circa 1946.

———, *Tank Gunnery*, Study Number 53, circa 1946.

———, *The Tactical Air Force in the European Theater of Operations*, Study Number 54, circa 1946.

Gibbons, Phil, "The Urban Area During Stability Missions Case Study: East Timor," in Glenn, ed., *Capital Preservation*, 2001.

Gibler, Michael L., "Stryker Brigade Combat Team: Organization and Employment," briefing, 2006.

Gillie, Mildred, *Forging the Thunderbolt: A History of the Development of the Armored Force*, Harrisburg, Pa.: Military Service Publishing Company, 1947.

Glantz, David M., *The Soviet Airborne Experience*, Research Survey No. 4, Fort Leavenworth, Kan.: Combat Studies Institute, 1984.

Glenn, Russell W., *Reading Athena's Dance Card: Men Against Fire in Vietnam*, Annapolis: Naval Institute Press, 2000.

——— ed., *Capital Preservation: Preparing for Urban Operations in the Twenty-First Century: Proceedings of the RAND Arroyo-TRADOC-MCWL-OSD Urban Operations Conference, March 22–23, 2000*, Santa Monica, Calif.: RAND Corporation, CF-162-A, 2001. As of February 5, 2008:
http://www.rand.org/pubs/conf_proceedings/CF162/

Gonzales, Daniel, Michael Johnson, Jimmie McEver, Dennis Leedom, Gina Kingston, and Michael S. Tseng, *Network-Centric Operations Case Study: The*

Stryker Brigade Combat Team, Santa Monica, Calif.: RAND Corporation, MG-267-1-OSD, 2005. As of February 5, 2008:
http://www.rand.org/pubs/monographs/MG267-1/

Gordon, John, IV, David E. Johnson, and Peter A. Wilson, "Air-Mechanization: An Expensive and Fragile Concept," *Military Review*, Vol. 87, No. 1, January–February 2007, pp. 63–73.

Gordon, John, IV, and Jerry Sollinger, "The Army's Dilemma," *Parameters*, Vol. 34, No. 2, Summer 2004, pp. 33–45.

Gordon, Michael R., and Bernard E. Trainor, *The Generals' War: The Inside Story of the Conflict in the Gulf*, Boston: Little, Brown and Company, 1995.

Gourley, Scott R., "Stryker Scores with US Tactical Vehicle Force," *Jane's International Defence Review*, June 1, 2006.

Grant, Greg, "Army 'Reset' Bill Hits $9 Billion: Nearly 1,000 Vehicles Lost in Combat," *Army Times*, February 20, 2006, p. 16.

Grau, Lester M., "Russian Urban Tactics: Lessons from the Battle for Grozny," *National Defense University Strategic Forum 38*, 1994. As of February 5, 2008:
http://handle.dtic.mil/100.2/ADA394517

———, ed., *The Bear Went Over the Mountain: Soviet Combat Tactics in Afghanistan*, Washington, D.C.: National Defense University Press, 1996.

———, "Road Warriors of the Hindu Kush: The Battle for the Lines of Communication in the Soviet-Afghan War, Fort Leavenworth, Kan.: Foreign Military Studies Office, August 1996.

———, "Russian-Manufactured Armored Vehicle Vulnerability In Urban Combat: The Chechnya Experience," *Red Thrust Star*, January 1997. As of February 5, 2008:
http://www.fas.org/man/dod-101/sys/land/row/rusav.htm

———, "A Weapon for all Seasons: The Old But Effective RPG-7 Promises to Haunt the Battlefields of Tomorrow," Fort Leavenworth, Kan.: Foreign Military Studies Office, 1998. As of February 5, 2008:
http://leav-www.army.mil/fmso/documents/weapon.htm

———, "Mine Warfare and Counterinsurgency: The Russian View," *Engineer*, Vol. 29, No. 1, March 1999, pp. 2–6. As of February 5, 2008:
http://leav-www.army.mil/fmso/documents/minewar/minewar.htm

———, "Technology and the Second Chechen Campaign: Not All New and Not That Much," in Aldis, ed., 2000.

———, and Ali Ahmad Jalali, "Underground Combat, Stereophonic Blasting, Tunnel Rats and the Soviet-Afghan War," *Engineer*, Vol. 28, No. 4, November

1998, pp. 20–23. As of February 5, 2008:
http://leav-www.army.mil/fmso/documents/undrgrnd/undrgrnd.htm

Grau, Lester M., and William A. Jorgenson, and Robert R. Love, "Guerilla Warfare and Land Mine Casualties Remain Inseparable," *U.S. Army Medical Department Journal*, Vol. PB 8-98-10/11/12, October–December 1998, pp. 10–16. As of February 5, 2008:
http://leav-www.army.mil/fmso/documents/guerwf/guerwf.htm

Grau, Lester M., and Jacob W. Kipp, "Urban Combat: Confronting the Specter," *Military Review*, Vol. 79, No. 4, July–August 1999, pp. 9–17. As of February 5, 2008:
http://leav-www.army.mil/fmso/documents/urbancombat/urbancombat.htm

Grau, Lester M., and Timothy Smith, "A 'Crushing' Victory: Fuel-Air Explosives and Grozny 2000," *Marine Corps Gazette*, Vol. 84, No. 8, August 2000, pp. 30–33. As of February 5, 2008:
http://leav-www.army.mil/fmso/documents/fuelair/fuelair.htm

Grau, Lester M., and Timothy L. Thomas, "Russian Lessons Learned from the Battles for Grozny," *Marine Corps Gazette*, Vol. 84, No. 4, April 2000, pp. 45–48. As of February 5, 2008:
http://leav-www.army.mil/fmso/documents/Rusn_leslrn.htm

Grau, Lester M., and Mohammand Yahya, "The Soviet Experience in Afghanistan," *Military Review*, Vol. 75, No. 5, October 1995, pp. 16–27.

Green, Constance McLaughlin, Harry C. Thomson, and Peter C. Roots, *The Ordnance Department: Planning Munitions for War*, Washington, D.C.: Office of the Chief of Military History, Department of the Army, [1955] 1990.

"The Gulf War: A Chronology," *Air Force*, Vol. 84, No.1, January 2001. As of February 5, 2008:
http://www.afa.org/magazine/jan2001/0101chrono.asp

Haight, David B., "Operation JUST CAUSE: Foreshadowing Example of Joint Vision 2010 Concepts in Practice," thesis, Newport, R.I.: Naval War College, 1998.

Hallanan, George H., Jr., "The Go-Anywhere Tank Company: The 603rd in the Southwest Pacific," *Army*, Vol. 41, No. 1, September 1994, pp. 51–54.

Hammond, Kevin J., and Frank Sherman, "Sheridans in Panama," *Armor*, Vol. 99, No. 2, March–April 1990, pp. 8–15.

Harrison, Gordon A., *Cross-Channel Attack*, Washington, D.C.: Center of Military History, U.S. Army, [1951] 1984.

Hart, S., and R. Hart, *German Tanks of World War II*, New York: Barnes and Noble, 1999.

Hastings, Max, *Armageddon: The Battle for Germany, 1944–1945*, New York: A. A. Knopf, 2004.

Heitman, Helmoed-Romer, "Operations Moduler and Hooper," in A. de la Rey, ed., *South African Defence Review*, Durban, South Africa: Walker Ramus, pp. 275–294. As of February 5, 2008:
http://www.rhodesia.nl/modhoop.htm

———, *War in Angola: The Final South African Phase*, Gibraltar: Ashanti, 1990.

Herbst, Jeffrey, "Prospects for Revolution in South Africa," *Political Science Quarterly*, Vol. 103, No. 4, Winter 1988–1989, pp. 665–685.

Herring, George C., *America's Longest War: The United States and Vietnam, 1950–1975*, 3rd ed., New York: McGraw-Hill, 1996.

Hines, Jay A., "Confronting Continuing Challenges: A Brief History of the United States Central Command," n.d. As of July 26, 2001:
http://www.centcom.mil/what%20is/history.htm

Hirsch, John L., and Robert B. Oakley, *Somalia and Operation Restore Hope: Reflections on Peacemaking and Peacekeeping*, Washington, D.C.: United States Institute of Peace, 1995.

"History of the 10th Mountain Division," U.S. Army Installation Management Agency, Northeast Region, October 22, 2004. As of February 5, 2008:
http://www.drum.army.mil/sites/about/hist-10mtn.asp

History Office, XVIII Airborne Corps and Joint Task Force South, "Corps Historian's Personal Notes Recorded During the Operation," n.d. As of February 5, 2008:
http://www.army.mil/cmh-pg/documents/panama/notes.htm

———, "Operation Just Cause: List of Participating Units," n.d. As of February 5, 2008:
http://www.army.mil/cmh-pg/documents/panama/unitlst.htm

———, "Panamanian Defense Force Order of Battle: Operation Just Cause," n.d. As of February 5, 2008:
http://www.army.mil/cmh-pg/documents/panama/pdfob.htm

Hoar, Joseph P., "A CINC's Perspective," *Joint Force Quarterly*, No. 2, Autumn 1993, pp. 56–63.

Hobbs, Joseph P., *Dear General: Eisenhower's Wartime Letters To Marshall*, Baltimore: Johns Hopkins University Press, 1999.

Hofmann, George F., "The Demise of the U.S. Tank Corps and Medium Tank Development Program," *Military Affairs*, Vol. 37, No. 1, February 1973, pp. 20–25.

———, and Donn A. Starry, eds., *From Camp Colt to Desert Storm: The History of U.S. Armored Forces*, Lexington, Ky.: University of Kentucky Press, 1999.

Holley, I. B., *Technology and Military Doctrine: Essays on a Challenging Relationship*, Maxwell-Gunter Air Force Base, Ala.; Air University Press, 2004.

Hornback, Paul, "The Wheel Versus Track Dilemma," *Armor*, Vol. 107, No. 2, March–April 1998, pp. 33–34.

House, Jonathan M., *Combined Arms Warfare in the Twentieth Century*, Lawrence, Kan.: University of Kansas Press, 2001.

Houston, Donald E., *Hell on Wheels: The 2d Armored Division*, Novato, Calif.: Presidio Press, 1986.

Hughes, Thomas Alexander, *Over Lord: General Pete Quesada and the Triumph of Tactical Air Power in World War II*, New York: Free Press, 1995.

Hunnicutt, R. P., *Stuart: A History of the American Light Tank*, Vol. 1, Novato, Calif.: Presidio Press, 1992.

———, *Sherman: A History of the American Medium Tank*, Novato, Calif.: Presidio Press, 1994.

Institute for National Strategic Studies, *Strategic Assessment 1996: Instruments of U.S. Power*, Washington, D.C.: National Defense University Press, 1996.

International Institute for Strategic Studies, *The Military Balance: 1989–1990*, London: Brassey's, 1989.

Irzyk, Albin F., *He Rode Up Front for Patton*, Raleigh, N.C.: Pentland Press, 1996.

Jane's Armour and Artillery, 2006, online edition.

Jarymowycz, Roman Johann, *Tank Tactics: From Normandy to Lorraine*, Boulder, Colo.: Lynne Rienner, 2001.

Jaster, Robert, *The 1988 Peace Accords and the Future of South-western Africa*, Adelphi Paper 253, London: International Institute for Strategic Studies, 1990.

Jean, Grace, "Stryker Units Win Over Skeptics," *National Defense*, Vol. XC, No. 623, October 2005, pp. 30–35.

Jensen, Marvin G., "An Independent Tank Battalion in World War II: How It Was Used . . . And Sometimes Misused," *Armor*, Vol. 108, No. 3, May–June 1999, pp. 27–28, 42.

Jentz, Thomas L., ed., *Panzertruppen: The Complete Guide to the Creation and Employment of Germany's Tank Force, 1933–1942*, Atglen, Pa.: Schiffer, 1996.

———, *Panzertruppen 2: The Complete Guide to the Creation and Combat Employment of Germany's Tank Force, 1943–1945*, Atglen, Pa.: Schiffer, 1996.

Johnson, Douglas V., II, ed., *Warriors in Peace Operations*, Carlisle Barracks, Pa.: Strategic Studies Institute, 1999.

Johnson, David E., *Fast Tanks and Heavy Bombers: Innovation in the U.S. Army, 1917–1954*, Ithaca: Cornell University Press, 1998.

Johnson, Wendell G., "The Employment of Supporting Arms in the Spanish Civil War," *C. & G.S.S. Quarterly*, Vol. 19, No. 72, March 1939, pp. 5–21.

Keegan, John, *The Illustrated Face of Battle*, New York: Viking, 1988.

Kelly, Michael J., Timothy L. H. McCormack, Paul Muggleton, and Bruce M. Oswald, "Legal Aspects of Australia's Involvement in the International Force for East Timor," *International Review of the Red Cross*, No. 841, March 31, 2001, pp. 101–139. As of February 5, 2008:
http://www.icrc.org/web/eng/siteeng0.nsf/html/57JQZ2

Kenety, Brian, "MPs Agree on Compensation for Victims of 1968 Soviet-Led Invasion," *Czech Radio*, February 25, 2002. As of February 5, 2008:
http://www.radio.cz/en/article/63804

Khodarenok, Mikhail, "Rukovodit' Operatziyey Porucheno Chekistam [Control of Operation Assigned to Chekhists]," *Nezavisimoye Voyennoye Obozreniye* (Internet edition), January 26, 2001. As of February 5, 2008:
http://nvo.ng.ru/wars/2001-01-26/1_operation.html

Kinnear, James, "Russian Armour Developments—Airborne AFVs (1950–93)," *Jane's Intelligence Review*, March 1994, pp. 109–113.

Korbut, Andrei, "Uchyoba V Boyu [Learning by Battle]," *Nezavisimoye Voyennoye Obozreniye*, December 24, 1999.

Korotchenko, Igor', "Brontekhnika Shtormuyet Gori [Armor Storms the Mountains]," *Nezavisimoye Voyennoye Obozreniye*, September 17, 1999.

Krause, Michael, Lieutenant (Australian Army, 2nd Cavalry Regiment), discussion with Fred Bowden, Australian Liaison at RAND Corporation, May 9, 2002.

Kulikov, Anatoly S., "Russian Internal Troops and Security Challenges in the 1990s," trans. R. Love, *Low-Intensity Conflict & Law Enforcement*, Vol. 3, No. 2, Autumn 1994. As of February 5, 2008:
http://www.fas.org/irp/world/russia/mvd/rusinttr.htm

———, "Trouble in the North Caucasus," *Military Review*, Vol. 79, No. 4, July–August 1999, pp. 29–39. As of February 5, 2008:
http://leav-www.army.mil/fmso/documents/trouble/trouble.htm

———, "The Chechen Operation," in Glenn, ed., *Capital Preservation*, 2001.

———, "The First Battle of Grozny," in Glenn, ed., *Capital Preservation*, 2001.

Laity, Mark, "Timor: The Military Challenge," BBC Online Network, September 24, 1999. As of February 5, 2008:
http://news.bbc.co.uk/hi/english/world/asia-pacific/newsid_452000/452230.stm

LaPorte, Leon J., and MaryAnn B. Cummings, "Prompt Deterrence: The Army in Kuwait," *Military Review*, Vol. 77, No. 6, November–December 1997, pp. 39–44.

Laur, Timothy M., and Steven L. Llanso, *Encyclopedia of Modern U.S. Military Weapons*, New York: Berkley Books, 1995.

Lehner, Charles, "Light Enough to Get There, Heavy Enough to Win," *Armor*, Vol. 103, No. 4, July–August 1994, pp. 10–14.

Lenehan, John R., "The Impact of the White Paper on Australian Armour," *Defender*, Vol. VXIII, No. II, Winter 2001, pp. 5–9.

Lewy, Guenter, *America in Vietnam*, New York: Oxford University Press, 1978.

Lieven, Anatol, *Chechnya: Tombstone of Russian Power*, New Haven: Yale University Press, 1998.

Lord, R. S., "Operation Askari: A Sub-Commanders Retrospective View of the Operation," *Military History Journal of the South African Defense Force,* Vol. 22, No. 4, 1992. As of August 21, 2002:
http://www.geocities.com/Yosemite/Forest/1771/askari.htm

Loza, Dmitry, "How Soviets Fought in U.S. Shermans," trans. James F. Gebhardt, *Armor*, Vol. 105, No. 4, July–August 1996, pp. 21–31.

Lozano, Frank, "Balkan Report III: The Six-Bradley Scout Platoon in Bosnia," *Armor*, Vol. 105, No. 5, September–October 1996, pp. 26–37.

Luedeke, Kirk A., "Death on the Highway: The Destruction of Groupement Mobile 100," *Armor*, Vol. 110, No. 1, January–February 2001, pp. 22–29.

Lynn, Adam, "Back to Iraq, with Tears and Courage," *News Tribune* (Tacoma, Wash.), June 3, 2006. As of February 5, 2008:
http://dwb.thenewstribune.com/news/military/stryker/story/5786857p-5171341c.html

MacDonald, Charles B., *The Last Offensive*, Washington, D.C.: Office of the Chief of Military History, U.S. Army, [1973] 1984.

Macksey, Kenneth, *Tank Versus Tank: The Illustrated Story of Armored Battlefield Conflict in the Twentieth Century*, Topsfield, Mass.: Salem House, 1988.

———, and John Batchelor, *Tank: A History of the Armoured Fighting Vehicle*, New York: Charles Scribner's Sons, 1970.

Magnuson, Stew, "Future Combat Vehicles Will Fall Short of Preferred Weight," *National Defense*, Vol. XCI, No. 643, June 2007, pp. 16–17. As of February 5, 2008:
http://www.nationaldefensemagazine.org/issues/2007/June/FutureCombat.htm

Mahler, Michael D., *Tinged in Steel: Armored Cavalry in Vietnam, 1967–69*, Novato, Calif.: Presidio Press, 1986.

Maksakov, Il'ya, "Federal'niye Voyska Prodolzhayut Nastupat' [Federal Forces Continue Attack]," *Nezavisimaya Gazeta*, January 13, 2000.

Malaysian Armed Forces, "United Nations Operation in Somalia II (1993–March 1995)," n.d. As of July 24, 2001: http://maf.mod.gov.my/english.atm/pbbdarat/unisom.html

Maloney, Sean M., "Insights into Canadian Peacekeeping Doctrine," *Military Review*, Vol. 76, No. 2, March–April 1996, pp. 12–23.

Maren, Michael, *The Road to Hell: The Ravaging Effects of Foreign Aid and International Charity*, New York: Free Press, 1997.

Matheny, Michael R., "Armor in Low-Intensity Conflict: The U.S. Experience in Vietnam," *Armor*, Vol. 97, No. 4, July–August 1988, pp. 9–15.

———, "Armor in Low-Intensity Conflict: The Soviet Experience in Afghanistan," *Armor*, Vol. 97, No. 5, September–October 1988, pp. 6–11.

Matloff, Maurice, ed., *American Military History*, rev. ed., Washington: Office of the Chief of Military History, U.S. Army, 1989.

Matyash, Vladimir, "Uchast' Banditov V Gorakh Predpreshena [Fate of Bandits in Mountains Predetermined]," *Kraznaya Zvezda*, February 9, 2000.

Maxwell, James J., "LAI: Impressions from SWA," *Marine Corps Gazette*, Vol. 75, No. 8, August 1991, pp. 18–19.

Mayo, Lida, *The Ordnance Department: On Beachhead and Battlefront*, Washington D.C.: Office of the Chief of Military History, U.S. Army, 1968.

Mazarr, Michael, *Light Forces and the Future of U.S. Military Strategy*, Washington, D.C.: Brassey's, 1990.

———, "Middleweight Forces for Contingency Operations," *Military Review*, Vol. 71, No. 8, August 1991, pp. 32–39.

McGrath, John J., ed., *An Army at War: Change in the Midst of Conflict*, Fort Leavenworth, Kan: Combat Studies Institute Press, 2006.

Melon, Charles D., Evelyn A. Englander, and David A. Dawson, *U.S. Marines in the Persian Gulf, 1990–1991: Anthology and Annotated Bibliography*, Washington, D.C.: History and Museums Division, Headquarters, U.S. Marine Corps, [1992] 1995.

Michaels, G. J., *Tip of the Spear: U.S. Marine Light Armor in the Gulf War*, Annapolis: Naval Institute Press, 1990.

Mikhailov, Andrei, "They've Learned to Use Tanks," *Nezavisimaya Gazeta*, May 25, 2000.

Millet Allan R., and Williamson Murray, eds., *The Interwar Period*, Boston: Unwin Hyman, 1990.

———, *Military Innovation in the Interwar Period*, New York: Cambridge University Press, 1996.

Minter, William, *King Solomon's Mines Revisited*, New York: Basic Books, 1986.

Morton, Matthew D., "Balkan Report II: Off-the-Shelf Wheeled Armor Proves its Worth in Macedonian Winter," *Armor*, Vol. 105, No. 4, July–August 1996, pp. 7–10.

Mudd, J. L., "Development of the American Tank-Infantry Team During World War II in Africa and Europe," *Armor*, Vol. 108, No. 5, September–October 1999, pp. 15–22.

"Na Argunsoye Ushchel'ye Sbrasivayut Ob'yemno-Detoniruyushchiye Bombi [Fuel-Air Bombs Being Dropped on Argun Region]," Lenta.ru, February 9, 2000. As of February 5, 2008:
http://www.lenta.ru/vojna/2000/02/09/argun/

Nagl, John A., *Learning to Eat Soup with a Knife*, rev. ed., Chicago: The University of Chicago Press, 2005.

National Academy of Sciences, *Technology for the United States Navy and Marine Corps, 2000–2035: Becoming a 21st Century Force*, Washington, D.C., 1997. As of February 5, 2008:
http://www.nap.edu/html/tech_21st/ovindex.htm#Contents

Neller, Robert B., "Marines in Panama: 1998–1990," research paper, U.S. Marine Corps Command and Staff College, 1991. As of February 5, 2008:
http://www.globalsecurity.org/military/library/report/1991/NRB.htm

New Zealand Army, "East Timor," n.d. As of June 11, 2001:
http://www.army.mil.nz/ops/easttimor_now.cfm

———, "Frequently Asked Questions: Light Armoured Vehicle (LAV III)," n.d. As of July 25, 2002:
http://www.army.mil.nz/nzarmy/grids/b_grid.asp?id=240&area=25

Newell, John F., III, "Airpower and the Battle of Khafji: Setting the Record Straight," thesis, Maxwell-Gunter Air Force Base, Montgomery, Ala.: School of Advanced Airpower Studies, Air University, 1998.

Nikolaev, Dmitri, "Voyska Idut V Gori [Troops Going to Mountains]," *Nezavisimoye Voyennoye Obozreniye*, February 11, 2000.

———, "Spetzoperatziya Zavershilas', a Voyna Prodolzhayetsya [Special Operation Complete, But War Continues]," *Nezavisimoye Voyennoye Obozreniye*, March 3, 2000.

Norval, Morgan, *Death in the Desert: The Namibian Tragedy*, Washington, D.C.: Selous Foundation Press, 1989. As of February 5, 2008:
http://www.geocities.com/Yosemite/Forest/1771/norch13.htm

Novichkov, N. N., et al., *Rossiiskiye Voenniye Sili V Chechenskom Konflikte: Analiz, Itogi, Vivody* [*The Russian Armed Forces in the Chechen Conflict: Analysis, Results, Conclusions*], Paris, Moscow: Kholveg-Infoglob, Trivola, 1995.

Nulsen, Charles K., "Advising as a Prelude to Command," essay, Carlisle, Pa.: U.S. Army War College, December 2, 1969.

Odom, William O., *After the Trenches: The Transformation of U.S. Army Doctrine, 1918–1939*, College Station, Tex.: Texas A&M University Press, 1999.

Office of the Controller and Auditor General, New Zealand, "New Zealand Defence Force: Deployment to East Timor," n.d. As of July 25, 2002:
http://www.oag.govt.nz/HomePageFolders/Publications/EastTimor/East_Timor.htm

Ogorkiewicz, R. M., "Rooikat 105: A New South African Combat Reconnaissance Vehicle," *Jane's International Defense Review*, March 1994, pp. 51–55.

———, "Achzarit: A Radically Different Armoured Infantry Vehicle," *Jane's International Defense Review*, September 1995, pp. 73–77.

———, "Technical Dossier: Infantry Armored Vehicle Design Continues to Vary," *Jane's International Defense Review*, August 1997, pp. 63–80.

Oliker, Olga, *Russia's Chechen Wars, 1994–2000: Lessons from Urban Combat*, Santa Monica, Calif.: RAND Corporation, MR-1289-A, 2001. As of February 5, 2008:
http://www.rand.org/pubs/monograph_reports/MR1289/

Orr, Michael, "Better or Just Not So Bad? An Evaluation of Russian Combat Effectiveness in the Second Chechen War," in Aldis, 2000.

"Osnovniye Boyi V Chechenskikh Gorakh Razvernutsya Cherez Dva Dnya [Main Battles in Chechen Mountains To Begin in Two Days]," Lenta.ru, February 9, 2000. As of February 5, 2008:
www.lenta.ru/vojna/2000/02/09/faza/

Palmer, Bruce, *The 25-Year War: America's Military Role in Vietnam*, New York: Touchstone, 1984.

Parliament of Australia, Joint Standing Committee on Foreign Affairs, Defence and Trade, "Technology, Equipment and Supplies," in *From Phantom to Force: Towards a More Efficient and Effective Army*, September 4, 2000, Chapter Eight. As of February 8, 2004:
http://www.aph.gov.au/house/committee/jfadt/army/Armych8.htm

Partridge, Ira L., "Deployable Versus Survivable," *Armor*, Vol. 110, No. 2, March–April 2001, pp. 12–14, 44.

Pengelley, Rupert, "Piranha: Mainstay of Medium Force Mobility," *Jane's International Defense Review*, April 2000, pp. 27–33.

Perry, Charles P., "Mogadishu, October 1993: A Company XO's Notes on Lessons Learned," *Infantry Magazine*, Vol. 84, No. 6, November–December 1994, pp. 31–38.

Phillips, R. Cody, *Operation Just Cause: The Incursion into Panama*, Washington, D.C.: Office of the Chief of Military History, U.S. Army, n.d.

Phipps, David M., "Yugoslavian Armor Fleet Is a Mix of New and (Some Very) Old," *Armor*, Vol. 108, No. 3, May–June 1999, pp. 18–19.

Pirnie, Bruce R., and Corazon M. Francisco, *Assessing Requirements for Peacekeeping, Humanitarian Assistance, and Disaster Relief*, Santa Monica, Calif.: RAND Corporation, MR-951-OSD, 1998. As of February 5, 2008: http://www.rand.org/pubs/monograph_reports/MR951/

Plummer, Anne, "Army Chief Tells President Restructuring Force Could Cost $20 Billion," *Inside the Army*, February 9, 2004.

Quinlivan, James, "Burden of Victory: The Painful Arithmetic of Stability Operations," *RAND Review*, Vol. 27, No. 2, Summer 2003, pp. 28–29.

Race, Jeffrey, *War Comes to Long An: Revolutionary Conflict in a Vietnamese Province*, Berkeley: University of California Press, 1972.

Raines, Edgar F., Jr., *Eyes of Artillery: The Origins of Modern U.S. Army Aviation in World War II*, Washington, D.C.: Center of Military History, 2000.

Reardon, Mark J., and Jeffery A. Charlston, *From Transformation to Combat: The First Stryker Brigade at War*, Washington, D.C.: Center of Military History, U.S. Army, 2007.

"Rebels Attack 2 Army Trains in Fierce Fight in Chechnya," *New York Times*, February 11, 2000.

Record, Jeffrey, *The Wrong War: Why We Lost Vietnam*, Annapolis: Naval Institute Press, 1998.

Reiff, Jack, "Brigade Combat Team Program Update," briefing, 7th International Artillery and Indirect Fire Symposium and Exhibition, March 21, 2002.

Reynolds, Nicholas E., *Just Cause: Marine Operations in Panama 1998–1990*, Washington, D.C.: Headquarters, U.S. Marine Corps, 1996.

Ricks, Thomas E., *Fiasco: The American Military Adventure in Iraq*, New York: The Penguin Press, 2006.

———, and Roberto Suro, "The Wheels Turn in Army Strategy," *The Washington Post*, November 16, 2000.

Rogers, Bernard William, *Cedar Falls–Junction City: A Turning Point*, Washington, D.C.: Department of the Army, 1989.

Romanyuk, Aleksandr, "Zakhvat Perevala [Capturing a Mountain Pass]," *Aremeiskii Sbornik*, December 2000, pp. 37–39.

Rose, Michael, "A Liddell Hart Approach to Peacekeeping," Liddell Hart Centre for Military Archives, King's College, London, 1999. As of February 5, 2008:
http://www.kcl.ac.uk/lhcma/info/lec99.htm

Rupe, Chad A., "The Battle of Grozny: Lessons for Military Operations on Urbanized Terrain," *Armor*, Vol. 108, No. 3, May–June 1999, pp. 20–23, 47.

"Russian Force Set to Start Operation in Chechen Mountains," *Jamestown Foundation Monitor*, Vol. VI, No. 29, February 10, 2000.

Ryan, Alan, *Primary Responsibilities and Primary Risks: Australian Defence Force Participation in the International Force East Timor*, Duntroon, Australia: Land Warfare Studies Centre, 2000.

Rybakov, Peter, "Foreign Military Digests—Soviet Versus German Tanks," *Military Review*, Vol. 25, No. 8, November 1945, pp. 118–126.

Scales, Robert H., Jr., *Firepower in Limited War*, Washington, D.C.: National Defense University Press, 1990.

Schreier, Konrad F., Jr., *Standard Guide to U.S. World War II Tanks and Artillery*, Iola, Wisc.: Krause Publications, 1994.

Schubert, Frank N., "U.S. Army Corps of Engineers and Afghanistan's Highways, 1960–1967," *Bridge to the Past*, No. 4, June 1996. As of February 5, 2008:
http://www.hq.usace.army.mil/history/bridge4.htm

———, and Theresa L. Kraus, eds., *The Whirlwind War*, Washington, D.C.: Center of Military History, U.S. Army, 1995.

Sheftick, Gary, "Army to Reset into Modular Brigade-Centric Force," Army News Service, February 24, 2004. As of February 5, 2008:
http://www4.army.mil/ocpa/read.php?story_id_key=5703

———, and Michele Hammonds, "Army Selects GM to Make Interim Armored Vehicles," Army News Service, November 20, 2000. As of February 5, 2008:
http://www.fas.org/man/dod-101/sys/land/docs/man-la-mav-001120.htm

Sherman, Frank, "Operation Just Cause: The Armor-Infantry Team In the Close Fight," *Armor*, Vol. 105, No. 5, September–October 1996, pp. 34–35.

Sherry, Michael S., *In the Shadow of War: The United States Since the 1930s*, New Haven: Yale University Press, 1995.

Shinseki, Erik K., "Address to the Eisenhower Luncheon," 45th Annual Meeting of the Association of the United States Army, October 12, 1999. As of February 5, 2008:

http://www.amc.army.mil/LOGCAP/docs/csavisionstatement.pdf#search=%22address%20to%20the%20eisenhower%20luncheon%2045th%20annual%22

Sliwa, Steven, "Maneuver and Other Missions in OIF: 1-37 FA, 3/2 SBCT," *Field Artillery*, No. PB6-05-2, March–April 2005, pp. 10–15.

Smithers, A. J., *A New Excalibur: The Development of the Tank, 1909–1939*, London: L. Cooper in association with Secker & Warburg, 1986.

Smolin, Andrei, and Viktor Kolomietz, "Khuzhe Gor Mogut Bit' Tol'ko Gori . . . [Only Mountains Are Worse Than Mountains . . .]," *Armeiskii Sbornik*, March 2000.

"Somalia: Operations Other Than War," Foreign Military Studies Office Special Study, No. 93-1, 1991.

Speyer, Arthur L., III, "The Two Sides of Grozny," in Glenn, ed., 2001.

Stanton, Shelby L., *Order of Battle: U.S. Army, World War II*, Novato, Calif.: Presidio Press, 1984.

Starry, Donn A., *Mounted Combat in Vietnam*, Washington, D.C.: Department of the Army, 1989.

Steadman, Kenneth A., *The Evolution of the Tank in the U.S. Army*, Fort Leavenworth, Kan.: Combat Studies Institute, U.S. Army Command and General Staff College, 1982. As of February 5, 2008:
http://cgsc.leavenworth.army.mil/carl/resources/csi/steadman2/steadman2.asp

Steele, Brett, *Military Reengineering Between the World Wars*, Santa Monica, Calif.: RAND Corporation, MG-253-OSD, 2005. As of February 5, 2008:
http://www.rand.org/pubs/monographs/MG253/

Steele, Robert D., "Intelligence Lessons Learned From Recent Expeditionary Operations," Command, Control, Communications, and Intelligence Department, Headquarters, U.S. Marine Corps, August 3, 1992. As of February 5, 2008:
www.oss.net/dynamaster/file_archive/040319/72c4a083bf9bc415e2501674658bccdb/OSS1999-P2-24.pdf

Steenkamp, Willem, *Borderstrike! South Africa into Angola*, Durban, South Africa: Butterworth, 1983.

Stevens, Roger M., and Kyle J. Marsh, "3/2 SBCT and the Countermortar Fight in Mosul," *Field Artillery*, No. PB6-05-1, January–February 2005, pp. 36–39.

Stiff, Peter, *The Silent War: South African Recce Operations, 1969–1994*, Cape Town: Galago Publishers, 1996.

Stratfor.com, "Recent Military Actions in Border Regions of Chechnya, Dagestan, Georgia Ending," n.d. As of September 14, 1999:
www.stratfor.com/CIS/countries/dagestanplot.htm

Stray, John E., "Mountain Warfare: The Russian Perspective," Foreign Military Studies Office, March 1994. As of February 5, 2008:
http://leav-www.army.mil/fmso/documents/mountain.htm

Sullivan, Brian R., "The Italian Armed Forces, 1918–1940," in Millet and Murray, eds., 1990.

———, "Fascist Italy's Military Involvement in the Spanish Civil War," *The Journal of Military History*, Vol. 59, No. 4, October 1995, pp. 697–727.

Taw, Jennifer Morrison, *Operation Just Cause: Lessons for Operations Other Than War*, Santa Monica, Calif.: RAND Corporation, MR-569-A, 1996. As of February 5, 2008:
http://www.rand.org/pubs/monograph_reports/MR569/

Thomas, Hugh, *The Spanish Civil War*, New York: Harper and Row, 1961.

Thomas, Timothy, "The Caucasus Conflict and Russian Security: The Russian Armed Forces Confront Chechnya: Military Activities of the Conflict During 11–31 December 1994," *The Journal of Slavic Military Studies*, Vol. 8, No. 2, June 1995, pp. 257–290. As of February 5, 2008:
http://www.globalsecurity.org/military/library/report/1995/chechpt2.htm

———, "The Caucasus Conflict and Russian Security: The Russian Armed Forces Confront Chechnya III: The Battle for Grozny, 1–26 January 1995," *The Journal of Slavic Military Studies*, Vol. 10, No. 1, March 1997, pp. 50–108. As of February 5, 2008:
http://www.globalsecurity.org/military/library/report/1995/chechpt3.htm

———, "Air Operations in Low Intensity Conflict: The Case of Chechnya," *Airpower Journal*, Vol. 11, No. 4, Winter, 1997, pp. 51–59. As of February 5, 2008:
http://www.globalsecurity.org/military/library/report/1997/thomas.htm

———, "The Battle of Grozny: Deadly Classroom for Urban Combat," *Parameters*, Vol. 29, No. 2, Summer 1999, pp. 87–102. As of February 5, 2008:
http://carlisle-www.army.mil/usawc/Parameters/99summer/thomas.htm

———, "Grozny 2000: Urban Lessons Learned," *Military Review*, Vol. 80, No. 4, July–August 2000, pp. 50–58. As of February 5, 2008:
http://leav-www.army.mil/fmso/documents/grozny2000/grozny2000.htm

———, "A Tale of Two Theaters: Russian Actions in Chechnya in 1994 and 1999," *Analysis of Current Events*, Vol. 12, Nos. 5–6, September 2000. As of February 5, 2008:
http://leav-www.army.mil/fmso/documents/chechtale.htm

Titus, James, *The Battle of Khafji: An Overview and Preliminary Analysis*, Maxwell-Gunter Air Force Base, Ala.: Air University, 1996.

"Transcript: Brig-Gen. Ham on the Stryker in Iraq," *Defense Industry Daily*, October 16, 2006. As of February 5, 2008:

http://www.defenseindustrydaily.com/transcript-briggen-ham-on-the-stryker-in-iraq-02698/

Travers, Tim, *The Killing Ground: The British Army, the Western Front, and the Emergence of Modern Warfare, 1900–1918*, London: Allen and Unwin, 1987.

Triplet, Robert H., *A Colonel in the Armored Divisions: A Memoir, 1941–1945*, Columbia. Mo.: University of Missouri Press, 2001.

Udmantzev, Vadim, "Polkoviye 'Zagogulini' [Regimental Bumbling]," *Nezavisimoye Voyennoye Obozreniye*, Internet edition, February 2, 2001. As of February 5, 2008:
http://nvo.ng.ru/notes/2001-02-02/8_shelfs.html

"The United Nations and East Timor: A Chronology," United Nations Peace and Security Web site, n.d. As of February 5, 2008:
http://www.un.org/peace/etimor99/chrono/body.html

The United National Security Council, *Resolution 1264*, S/RES/1264, September 15, 1999. As of February 5, 2008:
www.unmit.org/UNMISETwebsite.nsf/p9999/$FILE/1264.pdf

U.S. 12th Army Group, *12th Army Group Report of Operations (Final After Action Report)*, Vol. 11, *Antiaircraft Artillery, Armored, Artillery, Chemical Warfare and Signal Sections*, 1945.

U.S. Department of the Army, "Army Campaign Plan Briefing," n.d. As of September 19, 2004:
http://www.army.mil/thewayahead/acpdownloads.html

———, *Army Transformation Wargame 2001*, Headquarters, U.S. Army Training and Doctrine Command, Fort Monroe, Va., n.d.

———, "The United States Army Vision," n.d. As of September 19, 2001:
http://www.army.mil/vision/default.htm

———, *OPFOR Worldwide Equipment Guide*, U.S. Army Training and Doctrine Command, Deputy Chief of Staff for Intelligence, Fort Monroe, Va., 2001.

———, FM 3-0, *Operations*, 2001.

———, "Army Announces Name for Interim Armored Vehicle," press release, February 27, 2002.

———, "Joint Task Force South in Operation Just Cause, Oral History Interview JCIT 097Z (LTG Carmen Cavezza)," Fort Lewis, Wash., April 30, 1992. As of February 5, 2008:
http://www.history.army.mil/documents/panama/JCIT/JCIT97Z.htm

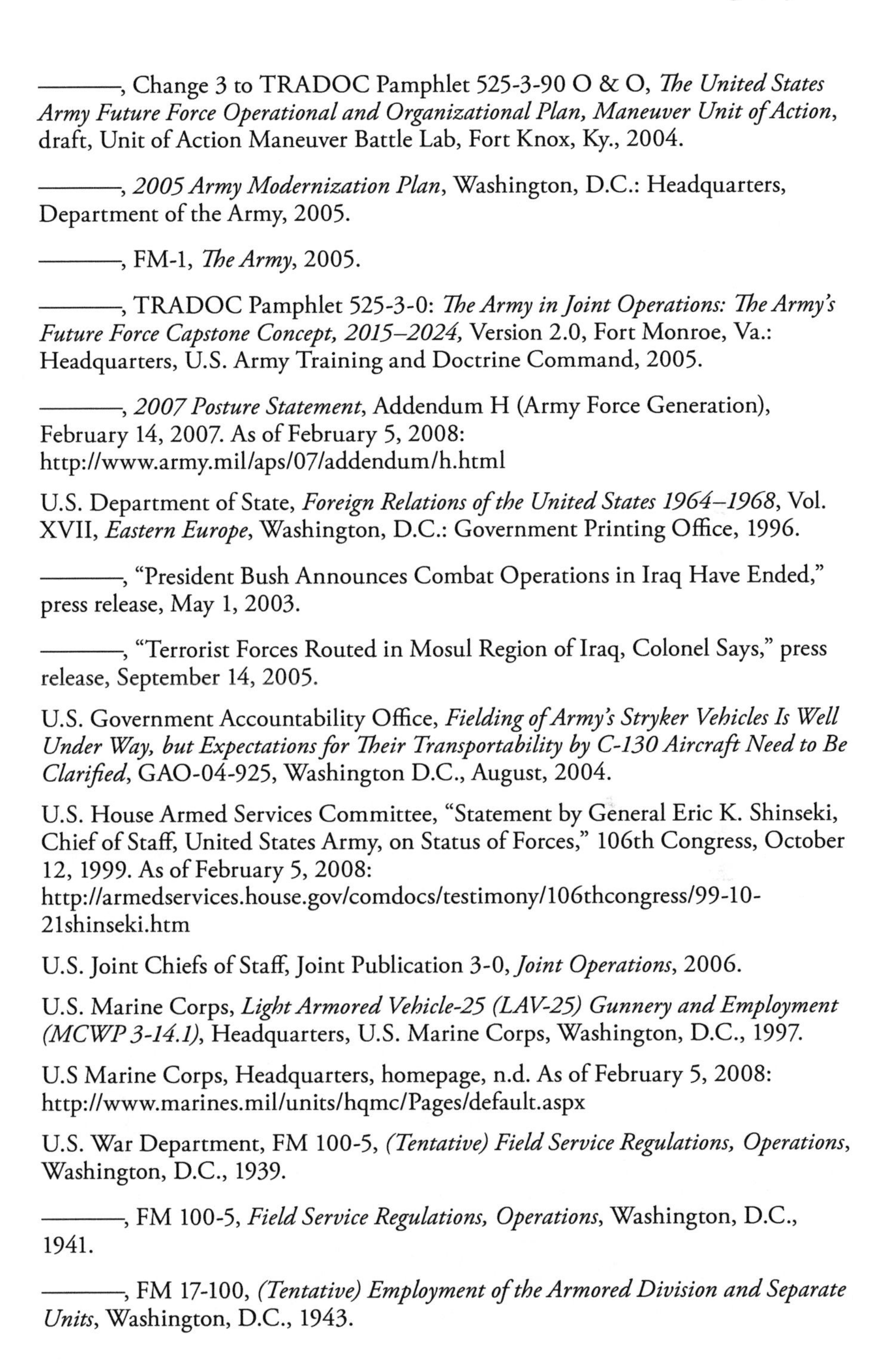

———, Change 3 to TRADOC Pamphlet 525-3-90 O & O, *The United States Army Future Force Operational and Organizational Plan, Maneuver Unit of Action*, draft, Unit of Action Maneuver Battle Lab, Fort Knox, Ky., 2004.

———, *2005 Army Modernization Plan*, Washington, D.C.: Headquarters, Department of the Army, 2005.

———, FM-1, *The Army*, 2005.

———, TRADOC Pamphlet 525-3-0: *The Army in Joint Operations: The Army's Future Force Capstone Concept, 2015–2024,* Version 2.0, Fort Monroe, Va.: Headquarters, U.S. Army Training and Doctrine Command, 2005.

———, *2007 Posture Statement*, Addendum H (Army Force Generation), February 14, 2007. As of February 5, 2008:
http://www.army.mil/aps/07/addendum/h.html

U.S. Department of State, *Foreign Relations of the United States 1964–1968*, Vol. XVII, *Eastern Europe*, Washington, D.C.: Government Printing Office, 1996.

———, "President Bush Announces Combat Operations in Iraq Have Ended," press release, May 1, 2003.

———, "Terrorist Forces Routed in Mosul Region of Iraq, Colonel Says," press release, September 14, 2005.

U.S. Government Accountability Office, *Fielding of Army's Stryker Vehicles Is Well Under Way, but Expectations for Their Transportability by C-130 Aircraft Need to Be Clarified*, GAO-04-925, Washington D.C., August, 2004.

U.S. House Armed Services Committee, "Statement by General Eric K. Shinseki, Chief of Staff, United States Army, on Status of Forces," 106th Congress, October 12, 1999. As of February 5, 2008:
http://armedservices.house.gov/comdocs/testimony/106thcongress/99-10-21shinseki.htm

U.S. Joint Chiefs of Staff, Joint Publication 3-0, *Joint Operations*, 2006.

U.S. Marine Corps, *Light Armored Vehicle-25 (LAV-25) Gunnery and Employment (MCWP 3-14.1)*, Headquarters, U.S. Marine Corps, Washington, D.C., 1997.

U.S Marine Corps, Headquarters, homepage, n.d. As of February 5, 2008:
http://www.marines.mil/units/hqmc/Pages/default.aspx

U.S. War Department, FM 100-5, *(Tentative) Field Service Regulations, Operations*, Washington, D.C., 1939.

———, FM 100-5, *Field Service Regulations, Operations*, Washington, D.C., 1941.

———, FM 17-100, *(Tentative) Employment of the Armored Division and Separate Units*, Washington, D.C., 1943.

———, FM 100-5, *Field Service Regulations, Operations*, Washington, D.C., 1944.

———, *Biennial Report of the Chief of Staff of the U.S. Army, General George C. Marshall, July 1, 1943, to June 30, 1945, to the Secretary of War*, Washington, D.C.: U.S. News Publishing Corp., 1945.

"V Gorniye Rayoni Chechni Perebrosheno Podkrepleniye [Reinforcements Sent to Mountainous Regions of Chechnya]," Lenta.ru, February 12, 2000. As of February 5, 2008:
www.lenta.ru/vojna/2000/02/12/help

Van Creveld, Martin, *Fighting Power: German and US Army Performance, 1939–1945*, Westport, Conn.: Greenwood Press, 1982.

Val'chenko, Sergey, and Konstantin Yur'yev, eds., "Krugliy Stol AS: Goryachiy Vozdukh Kavkaza [AS (Armeiskii Sbornik) Roundtable: Hot Air of the Caucasus]," *Armeiskii Sbornik*, February 2001, pp. 24–32.

Valenta, Jiri, "From Prague to Kabul: The Soviet Style of Invasion," *International Security*, Vol. 5, No. 2, Fall 1980, pp. 114–141.

Vigman, Fred K., "Eclipse of the Tank," *Military Affairs*, Vol. 8, No. 2, Summer 1944, pp. 101–108.

Viktorov, Andrei, "Predpraznichniy Shturm [Preholiday Storm]," *Segodnya*, February 24, 2000.

Walker, Jim, "Vietnam: Tanker's War?" *Armor*, Vol. 106, No. 3, May–June 1997, pp. 24–30.

Wallace, William S., "Victory Starts Here! Changing TRADOC to Meet the Needs of the Army," *Military Review*, Vol. 86, No. 3, May–June 2006, pp. 59–69.

Warford, James M., "Cold War Armor After Chechnya: An Assessment of the Russian T-80," *Armor*, Vol. 104, No. 36, November–December 1995, pp. 18–21.

Wathen, Julian, *Humanitarian Operations: The Dilemma of Intervention*, Occasional Paper Number 42, Shrivenham, UK: Strategic and Combat Studies Institute, 2001.

Weeks, John, *Men Against Tanks: A History of Anti-Tank Warfare*, New York: Mason/Charter, 1975.

Weigley, Russell F., *The American Way of War: A History of United States Military Strategy and Policy*, 2nd ed., Bloomington, Ind.: Indiana University Press, 1977.

———, *History of the United States Army*, enlarged ed., Bloomington, Ind.: Indiana University Press, 1984.

Westermann, Edward B., "The Limits of Soviet Airpower: The Bear Versus the Mujahideen in Afghanistan, 1979–1989," thesis, Maxwell-Gunter Air Force Base, Montgomery, Ala.: School of Advanced Airpower Studies, Air University, 1997.

Wilson, Dale E., *Treat 'Em Rough! The Birth of American Armor, 1917–1920*, Novato, Calif.: Presidio Press, 1989.

Wilson, John B., *Maneuver and Firepower: The Evolution of Divisions and Separate Brigades*, Washington, D.C.: Center of Military History, 1998.

———, "Organizing the First Armored Divisions," *Armor*, Vol. 108, No. 4, July–August 1999, pp. 41–43.

Wilson, Peter A., John Gordon IV, and David E. Johnson, "An Alternative Future Force: Building a Better Army," *Parameters*, Vol. 33, No. 4, Winter 2003–2004, pp. 19–31.

Winicki, Anthony, "The Marine Combined Arms Raid," *Marine Corps Gazette*, Vol. 75, No. 12, December 1991, pp. 54–55.

Winton, Harold R., *To Change an Army*, Lawrence, Kan.: University Press of Kansas, 1988.

———, and David R. Mets, eds., *The Challenge of Change: Military Institutions and New Realities, 1918–1941*, Lincoln, Neb.: University of Nebraska Press, 2000.

Womack, Scott, "AGS (Armored Gun System) in Low-Intensity Conflict: Flexibility Is the Key to Victory," *Armor*, Vol. 103, No. 2, March–April 1994, pp. 42–44.

Worrick, Christopher P., "The Battle of Suoi Tre: Viet Cong Infantry Attack on a Fire Base Ends in Slaughter When Armor Arrives," *Armor*, Vol. 109, No. 3, May–June 2000, pp. 23–28.

Zaloga, Steven J., "New Soviet Infantry Fighting Vehicle Features Tank-Like Armament," *Armed Forces Journal International*, Vol. 128, No. 2, September 1990, p. 25.

———, *Inside the Blue Berets: A Combat History of Soviet and Russian Airborne Forces, 1930–1995*, Novato, Calif.: Presidio Press, 1995.

———, "Soviet Tank Operations in the Spanish Civil War," *The Journal of Slavic Military Studies*, Vol. 12, No. 3, September 1999, pp. 134–162. As of February 5, 2008:
http://libraryautomation.com/nymas/soviet_tank_operations_in_the_sp.htm

———, *US Armored Divisions: The European Theater of Operations*, University Park, Ill.: Osprey Publishing, 2004.

Zetterling, Niklas, *Normandy 1944: German Military Organization, Combat Power and Organizational Effectiveness*, Winnipeg, Manitoba: J. J. Fedorowicz Publishing, 2000.

Zickel, Raymond E., *Soviet Union: A Country Study*, Washington, D.C.: Library of Congress, 1989. As of February 5, 2008:
http://lcweb2.loc.gov/frd/cs/sutoc.html

Ziemke, Earl F., "The Soviet Armed Forces in the Interwar Period," in Millet and Murray, eds., 1990.

Zumbro, Ralph, *The Iron Cavalry*, New York: Pocket Books, 1998.

Index